A TEXT BOOK OF

PRINCIPLES OF ELECTRICAL MACHINE DESIGN

FOR

SEMESTER – VI

THIRD YEAR (T.Y.) B. TECH COURSE IN ELECTRICAL ENGINEERING / ELECTRICAL ENGINEERING (ELECTRONICS AND POWER) / ELECTRICAL AND ELECTRONICS ENGINEERING / ELECTRICAL AND POWER ENGINEERING

Strictly According to New Revised Credit System Syllabus of Babasaheb Ambedkar Technological University (BATU), Lonere, (Dist. Raigad) Maharashtra, (w.e.f. June 2019-20)

LAXMIKANT V. BAGALE

ME (EMD)
Assistant Professor & Head
Electrical Engineering Department
M.B.E.S. College of Engineering,
AMBAJOGAI.

N1112

PRINCIPLES OF ELECTRICAL MACHINE DESIGNS (ELECT., BATU) ISBN : 978-93-89825-93-0

First Edition : January 2020
© : Author

Published By :
NIRALI PRAKASHAN
Abhyudaya Pragati, 1312 Shivaji Nagar
Off J.M. Road, PUNE 411005
Tel : (020) 25512336/37/39
Email : niralipune@pragationline.com

➢ DISTRIBUTION CENTRES

PUNE

Nirali Prakashan (Local) : 119 Budhwar Peth, Jogeshwari Mandir Lane, Pune 411002, Maharashtra
Tel : (020) 2445 2044, Mobile : 9657703145, Email : niralilocal@pragationline.com

Nirali Prakashan (Outstation) : S. No. 28/27 Dhayari, Near Asian College, Dhayari, Pune 411041, Maharashtra
Tel : (020) 2469 0204, Fax : (020) 2469 0316, Mobile : 9657703143
Email : bookorder@pragationline.com

MUMBAI

Nirali Prakashan : 385 S.V.P. Road, Rasadhara Co-op. Hsg. Society Ltd., Girgaum, Mumbai 400004, Maharashtra
Tel : (022) 2385 6339 / 2386 9976, Fax : (022) 2386 9976, Mobile : 9320129587
Email : niralimumbai@pragationline.com

➢ DISTRIBUTION BRANCHES

JALGAON

Nirali Prakashan : 34 V. V. Golani Market, Navi Peth, Jalgaon 425001, Maharashtra
Tel : (0257) 222 0395, Mob : 94234 91860, Email : niralijalgaon@pragationline.com

KOLHAPUR

Nirali Prakashan : New Mahadvar Road, Kedar Plaza 1st Floor, Opp. IDBI Bank
Kolhapur 416012, Maharashtra. Mobile : 9850046155
Email : niralikolhapur@pragationline.com

NAGPUR

Nirali Prakashan : Above Maratha Mandir, Shop No 3, Second Floor,
Rani Jhanshi Square, Sitabuldi, Nagpur 440012, Maharashtra
Tel : (0712) 254 7129, Email : niralinagpur@pragationline.com

DELHI

Nirali Prakashan : 4593/15 Basement, Agarwal Lane, Ansari Road, Daryaganj
Near Times of India Building, New DelhiV 110002 Mobile : 8505972553
Email : niralidelhi@pragationline.com

BENGALURU

Nirali Prakashan : Maitri Ground Floor, Jaya Apartments, No. 99, 6th Cross, 6th Main,
Malleswaram, Bengaluru 560003, Karnataka
Mobile : 9449043034, Email : niralibangalore@pragationline.com

niralipune@pragationline.com | www.pragationline.com
Also find us on ⓕ www.facebook.com/niralibooks

PREFACE

It gives me great pleasure to present the book **"Principles of Electrical Machine Design"** for the students of **Semester VI Third Year (T.Y.) B. Tech. Course Electrical Engineering / Electrical Engineering (Electronics and Power) / Electrical and Electronics Engineering / Electrical and Power Engineering of Dr. Babasaheb Ambedkar Technological University (BATU), Lonere, Dist. Raigad (Maharashtra).** This book is strictly as per the new revised syllabus 2019-20 Pattern, effective from the Academic Year July 2019-20.

In New Revised Syllabus, there will Class Assessment (CA) 20 Marks, Mid Sem. Exam. (MSE) 20 Marks and End Sem. Exam. (ESE) 60 Marks. End Sem. Exam. will be based on all Six units and each unit will carry 20 Marks.

The Theory Course will have 3 Credits.

The basic objective of this book is to bridge the gap between the vast contents of the reference books, written by the renowned International Authors and the concise requirements of Undergraduate Students. This book has been written in a comprehensive manner using Simple and Lucid language, keeping in mind students' requirements. The main emphasis has been given on exploring the basic concepts rather than merely the Information. Solved Examples, Summary and Exercises have been provided throughout the book and at the end of the Unit. Also I have given **Model Question Papers** for practice at the end of book.

My special thanks to my family members, students and all those who directly or indirectly supported us in this project.

I also take this opportunity to express my sincere thanks to Shri. Dineshbhai Furia, Shri. Jignesh Furia, Mrs. Nirali Verma, Shri. M. P. Munde and entire team of Nirali Prakashan, namely Mrs. Deepali Lachake (Co-ordinator), and her colleagues who really have taken keen interest and untiring efforts in publishing this text.

The advice and suggestions of my esteemed readers to improve the text are most welcome and will be highly appreciated.

Pune   **Author**

SYLLABUS

Unit I : Principles and Design of Electrical Machines   **6 Hrs**

Design of Electrical machines along with their parts and special features, rating, Specifications, Standards, Performance and other criteria to be considered, Brief study of magnetic, electric, dielectric and other materials, Introduction to machine design.

Unit II : Design of Electrical Apparatus   **6 Hrs**

Detailed design of heating coils, starters and regulators. Design of Electrical Devices Field coils, Chokes and lifting magnets.

Unit III : AC and DC Winding   **6 Hrs**

Types of dc windings, Pitches, Choice and design of simple/ duplex lap and wave winding, Concept of multiplex windings and reasons for choosing them, Single and double layer single phase AC winding with integral and fractional slots, Single and double layer integral and fractional slot windings of three phase. AC winding factors, Tests for fault finding in windings, Numerical examples.

Unit IV : Heating, Cooling and Ventilation   **6 Hrs**

Study of different modes of heat generation, Temperature rise and heat dissipation, Heating and Cooling cycles, heating and cooling time constants, their estimation, dependence and applications, Methods of cooling / ventilation of electrical apparatus, Thermal resistance, radiated heat quantity of cooling medium (Coolant) Numerical.

Unit V : Design of Transformer   **6 Hrs**

Design of distribution and power transformers, Types, Classification and specifications, Design and main dimensions of core, yoke, winding, tank (with or without cooling tubes) and cooling tubes, Estimation of leakage reactance, resistance of winding, No load current, Losses, Voltage regulation and efficiency, Mechanical force developed during short circuits, Their estimation and measures to counteract them, Testing of transformers as per I.S.S., Numerical examples.

Unit VI : Computer Aided Design of Electrical Machine   **6 Hrs**

Introduction, advantages various approaches of Computer Aided Designing, Computer Aided Designing of transformer, Winding of rotating Electrical Machines. Optimization of Design.

CONTENTS

D.C. ARMATURE WINDINGS

AC WINDINGS

PRINCIPLES AND DESIGN OF ELECTRICAL MACHINE

1.1 INTRODUCTION

- Electrical machine design involves application of science and technology to produce cost-effective, durable, quality and efficient machines. Also the machines should be designed as per standard specifications. The requirements like low cost and high quality will be conflicting in nature and so a compromise should be made between them.

- Aim of design is to determine the dimensions of each part of the machine, the material specification, prepare the drawings and furnish to manufacturing units. Design has to be carried out keeping in view the optimizing of the cost, volume and weight and at the same time achieving the desired performance as per specification. Knowledge of latest technological trends to supply a competitive product is a must. Design should conform to stipulations specified by International / National standards. Design is the most important activity.

- The designer should be familiar with the following aspects:

  - ➤ Thorough knowledge of international/national standards.

  - ➤ Properties of good electrical materials (like copper), magnetic materials (like silicon steels), insulating materials (like Epoxy mica), mechanical and metallurgical properties of all types of steel.

  - ➤ Governing laws of electrical circuits.

  - ➤ Laws of heat transfer.

  - ➤ Prices of materials used, foreign exchange rates, types of duties levied on products.

  - ➤ Labour rates of both ski lied and unskilled labour

  - ➤ Knowledge of competitor's products.

1.2 DESIGN OF MACHINE

Design is defined as a creative physical realization of theoretical concepts.

1.2.1 Engineering Design

Engineering Design is a application of science, technology and invention to produce machines to perform specified tasks with optimum economy and efficiency.

1.2.2 Factors for Consideration

The design should be carried out based on the given specification using available materials economically and to achieve the following three key factors.

1. Cost
2. Durability
3. Compliance with performance criteria as laid down in specifications.

1.3 DESIGN FACTORS

The mechanical force required for movement in rotating electrical machines can be produced both by

1. Electrostatic fields
2. Electromagnetic fields

Both the fields stores some energy

1. **Electrostatic Fields**

   In electrostatic machines, the energy density is limited by the dielectric strength of the medium used. For Air dielectric medium the energy density about $40J/m^3$.

   Voltages that can be developed and used by normal means, the forces produced by electrostatic effects are very weak.

2. **Electromagnetic Fields**

   In electromagnetic machines, magnetic effect is used for production of force and there is no comparable restriction in magnetic fields. Maximum value of flux density that can be used is about $1.6wb/m^2$.

   A small current can produce large mechanical force by electromagnetic means and therefore all the modern electrical machines are electromagnetic type

1.4 CONSTRUCTIONAL ELEMENTS OF TRANSFORMER

- The transformer is a static electromagnetic device used to transfer electrical energy from a high potential (voltage) circuit to low potential (voltage) circuit or vice-versa.

- It consists of two or more windings which link with a common magnetic field. An iron core serves as a path for magnetic flux.

- **The constructional elements of the transformer are windings, core, tank and cooling tubes or radiators.**

- A simple transformer has two windings and they are called high voltage winding and low voltage winding. One of the winding is connected to supply and it is called primary. The other winding is connected to load and it is called secondary.

- The two different types of transformer constructions are core type and shell type. In core type transformer the windings surround the core and in shell type transformer the core surround the windings. The core and winding assembly is housed in the tank. Cooling tubes or radiators are provided around the tank surface in order to increase the effective cooling surface.

1.5 CONSTRUCTIONAL ELEMENTS OF ROTATING MACHINE

- The rotating electrical machines converts electrical energy to mechanical energy or vice-versa. The energy conversion takes place through magnetic field. Every rotating machine have the following three quantities. The presence of any two quantities, will produce the third quantity.

  - ➢ Magnetic Field - (Field)
  - ➢ Magnetic field - (Armature)
  - ➢ Mechanical force

- In generator. the armature is rotated by a mechanical force inside a magnetic field or the magnetic field is rotated by keeping armature stationary. By Faraday's law of induction, an emf is induced in the armature. When the generator is loaded, the armature current flows, which produce another magnetic field (armature magnetic field). Hence in a generator, by the presence of a magnetic field and mechanical force, an another magnetic field is produced.

- The mechanical force developed by the motor is due to the reaction of two magnetic fields, A current carrying conductor has a magnetic field around it. When it is placed in another magnetic field it experiences a mechanical force due to the reaction of two magnetic field. Hence in a motor by the presence of two magnetic fields a mechanical force is developed.

- From the above discussion it is clear that any rotating machine requires two magnetic field and one of the field is rotating. Hence a rotating machine will have a stationary and rotating electromagnet, each consisting of a core and winding. The stationary electromagnet is called stator and the rotating electromagnet is called rotor.

- The basic constructional elements of a rotating electrical machine are stator and rotor. In dc machines the stator consists of field core and winding. The rotor comprises of armature core and winding. In ac machines the stator has armature core and winding. The rotor consists of field core and winding.

- The constructional elements of various electrical machines are listed here.

1. **Constructional Elements of DC Machine,**

Stator	Rotor
• Yoke or Frame	• Armature core
• Field winding	• Armature winding
• Poles	• Commutator
• Pole shoe	• Others Brush
• Field winding interpole	• Brush holder

2. **Constructional Elements of Salient Pole Synchronous Machine**

Stator	Rotor
• Frame	• Field pole,
• Armature core	• Pole shoe.
• Armature winding	• Field winding.
• Damper winding	

3. **Constructional Elements of Cylindrical Rotor Synchronous Machine**

Stator	Rotor
• Frame	• Solid rotor
• Armature core	• Field conductors or bars
• Armature winding	

4. **Constructional Elements of Squirrel Cage Induction Motor**

Stator	Rotor
• Frame	• Rotor core
• Stator core	• Rotor bars
• Stator winding	• End ring

5. **Constructional Elements of Slip-Ring Induction Motor**

Stator	Rotor
• Frame	• Rotor core
• Stator core	• Rotor winding
• Stator winding	• Slip ring

1.6 BASIC STRUCTURE OF ELECTRICAL MACHINE

The Basic Structure of Electrical Machines

It consists of following parts

1. Magnetic circuits
2. Electric circuits
3. Dielectric circuits
4. Thermal circuits
5. Mechanical parts

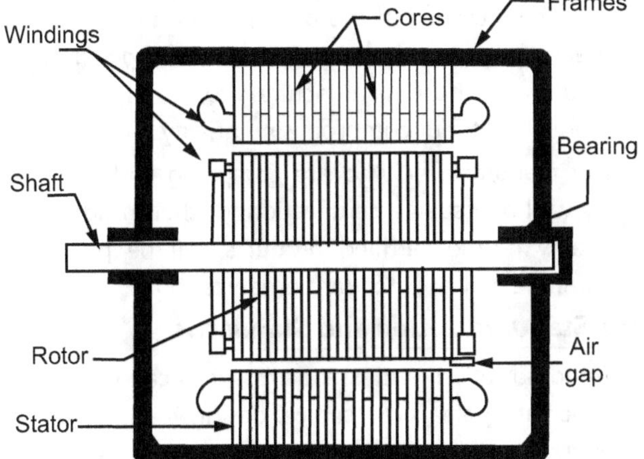

Fig. 1.1: Basic constructional details of rotating machines

1. **Magnetic Circuits :** It provides the path for the magnetic flux and consists of air gap, stator and rotor teeth and stator and rotor cores (yokes)

2. **Electric Circuits :** It consists of stator and rotor winding, winding of transformer. It should ensure that required emf is induced with no complexity in winding arrangement.

3. **Dielectric Circuits :** The dielectric circuit consists of insulation required to isolate one conductor from another and also the windings from the core. The insulating materials are essentially non-metallic and may be organic or inorganic ,natural or synthetic.

4. **Thermal Circuits :** The thermal circuits is concerned with mode and media for dissipation of heat produced inside the machine on account of losses.

5. **Main Mechanical Parts**

   - Frame
   - Bearings
   - Shaft

   A successful design brings out an economic compromise for space occupied by iron, copper (aluminum) insulation and coolant (which may be air, hydrogen, water or oil)

1.7 CLASSIFICATION OF DESIGN PROBLEMS

- The design of an electrical machine involves solution of many complex and engineering problems.

- **The Design Problems may be Classified under the Following Four Headings.**

  1. Electromagnetic design
  2. Mechanical design
  3. Thermal design
  4. Dielectric design

  Each problem may be solved Separately and the results are combined to give overall solution. Each of these three major problems may be further divided into simple problems and solutions of individual problem are combined to give the solution of a major design problem.

1. **Electromagnetic Design**

   The electromagnetic design problem in rotating machines involves the design of stator & rotor core dimensions, stator & rotor teeth dimensions, air-gap length, stator and rotor windings. In transformer, it is the problem of designing the core and the windings.

2. **Mechanical Design**

   The mechanical design in rotating machine involves the design of frame (enclosure), shaft and bearings. In transformer, it is the design of tank (i.e., housing for core and winding assembly).

3. **Thermal Design**

   The thermal design in rotating machine involves the design of cooling ducts in core and cooling fans. In case of large machines coolants like air or hydrogen may be forced to circulate in the ducts and air-gap. In transformer, it involves the design of cooling tubes or radiators.

4. **Dielectric Design**

   Another important design problem, that may require great attention is the design of insulations (Dielectric design). Dielectric materials are used to insulate one conductor from other and also the windings from the core. The dielectric materials are designed to withstand high voltage stresses. The breakdown of dielectric materials may lead to failure of machine.

1.8 LIMITATIONS IN DESIGN

Following are the limitations for design :

1. Saturation
2. Current Density
3. Temperature rise

4. Insulation
5. Efficiency
6. Mechanical parts
7. Commutation
8. Power factor
9. Consumers specifications
10. Standard specifications

1. **Saturation :** Electromagnetic machines use ferromagnetic materials .The maximum allowable flux density to be used is determined by the saturation level of the ferromagnetic material used. A high value of flux density results in increased core loss and increased excitation resulting in higher cost for the field system. It also introduces harmonics.

2. **Current Density :** Higher current density reduces the volume of copper but increases the losses and temperature.

3. **Temperature Rise :** The most important parts of the machine is insulation. The operating life of a machine depends upon the types of insulating materials used in its construction .The life of insulating materials in turn depends upon the temperature rise of the machine. Proper cooling and ventilation techniques are required to keep the temperature rise within safe limits.

4. **Insulation :** The insulation materials used in a machine should be able to withstand the electrical ,mechanical and thermal stresses which are produced in the machines. The type of insulation is decided by the maximum operating temperature of the machine parts where it is put. And also the size of the insulation is decided by maximum voltage stress and mechanical stresses produced.

5. **Efficiency :** The efficiency of the machine should be as high as possible to reduce the operating costs.

6. **Mechanical Parts :** The construction of mechanical parts should be as simple as possible and also it is technologically good. The design of mechanical part is particularly important in case of high speed machines.

7. **Commutation :** The problem of commutation is important in the case of commutator machines. Commutation condition limit the maximum output that can be taken from a machine.

8. **Power Factor :** Poor power factor results in larger values of current for the same power and therefore large conductor sizes have to be used. Power factor problem is particularly important in case of induction motor.

9. **Consumer's Specifications :** The specifications as laid down in the consumer's order have to be met and the design evolved should be such that it satisfied all the specifications and also the economic constraints imposed on the manufacturer.

10. **Standard Specifications :** This specifications are the biggest strain on the design because both the manufacturer as well as the consumer cannot get away from them without satisfying them.

1.9 STANDARD SPECIFICATIONS

- Every country has standard organization to fix standard specifications for the manufacturer. The specification is guidelines for the manufacturers to produce economic products without compromising quality.

- The manufacturer who are compiling with standards and will be issued a certification for their products. The quality of the certified products will be periodically monitored by the standard organization.

1.9.1 Standardization and Standards

- Standardization and standard specifications play an important part in the choice, design manufacture and operation of any apparatus.

- Every country has a standards organization to fix standard specifications for the manufacturers.

- The specification are guidelines for the manufacturers to produce economic products without compromising quality.

- The manufacturers who are compiling with the standards will be issued a certification for their products.

- The quality of the certified products will be periodically monitored by the standard organization.

Advantages of Standardization

- To manufacturer
  - ➢ Reduction in cost as economy results when number of objects are built at the same time.
  - ➢ Easy to production planning.
  - ➢ Stream-lining the a production line.
- To user
  - ➢ Standardization means interchangeability of equipments and spares.
- To designer
  - ➢ It means Rigidity

Indian standards these are issued by Bureau of Indian Standards (BIS), Manak Bhavan, Bahadur Shah Zafar Road. New Delhi

- **The Standard Specifications Issued for Electrical Machines :**
  - ➤ Standard ratings of machines
  - ➤ Types of enclosure
  - ➤ Standard dimensions of conductors to be used
  - ➤ Method of marking ratings and name plate details.
  - ➤ Performance specifications to be used
  - ➤ Types of insulation and permissible temperature rise
  - ➤ Permissible losses and range of efficiency
  - ➤ Procedure for testing of machine parts and machines
  - ➤ Auxiliary equipments to be provided
  - ➤ Cooling methods to be adopted
- In India, the Bureau of Indian standards (BIS) has laid down their specification (ISI) for various products. The standards will be amended time to time, in order to include the latest developments in technology.
- The name plate of the rotating machine has to bear the following details as per ISI specifications.
  - ➤ KW or KVA rating of machine
  - ➤ Rated working voltage
  - ➤ Operating speed
  - ➤ Full load current
  - ➤ Class of insulation
  - ➤ Frame size
  - ➤ Manufacturers name
  - ➤ Serial number of the product

1.9.2 Some of the Indian Standard Specifications Numbers along with Year of Issue for Electrical Machines are Listed Here

- IS 325 - 1996 : Specifications for three phase induction motor.
- IS 1231 - 1974 : Specifications for foot mounted induction motor.
- IS 4029 - 1967 : Guide for testing three phase induction motor.
- IS 996 - 1979 : Specifications for single phase cc and universal meter.
- IS 1885 - 1993 : Specifications for electric and magnetic circuits.
- IS 9499 - 1980 : Conventions concerning electric and magnetic circuits.
- IS 7538 - 1996 : Specifications of three phase induction motor centrifugal pumps and agricultural applications.
- IS 12615 - 1986 : Specifications for energy efficient induction motor.
- IS 9320 - 1979 Guide for testing dc machines.

- IS 4722 - 1992 : Specifications for rotating electrical machines.
- IS 12802 - 1989 : Temperature rise measurement of rotating electrical machines.
- IS 4889 - 1968 Method of determination of efficiency of rotating electrical machines.
- IS 13555 - 1993 : Guide for selection and application of three phase induction for different types of driven equipment.
- IS 7132 -1973 : Guide for testing Synchronous machines.
- IS 5422 - 1996 : Turbine type generators.
- IS 7572 - 1974 : Guide for testing single-phase ac and universal motors.
- IS 8789 - 1996 : Values of performance characteristics for three phase induction motors.
- IS 12066 - 1986 : Three phase induction motors for machine tools.
- IS 1180 - 1989 : Specifications for outdoor 3-phase distribution transformer 100 kVA. [Sealed and Non-sealed type)
- IS 2026 - 1994 : Specifications of power transformers.
- IS 11171 - 1985 :_ Dry type power transformers.
- IS 5142 - 1969 : Continuously variable voltage auto transformers
- IS 10028 -1985 : Code of practice for selection, installation and maintenance of transformers.
- IS 10561 - 1983 : Application guide for power transformers.
- IS 13956 - 1994 : Testing transformers.
- IS 9678 - 1980 : Methods of measuring temperature rise of electrical equipment.
- IS 12063 - 1987 : Classification of degree of protections provided by enclosures of electrical equipment.
- IS 3855 - 1966 : Standard dimensions of rectangular enamelled Copper conductor.
- IS 449 - 1962, : Standard dimensions of enamelled round Copper conductor (oleo resinous enamel).
- IS95 - 1960 : Standard dimensions of enamelled round copper conductor (synthetic enamel).
- IS 1897 - 1962: Standard dimensions of bore copper strip.
- IS 1666 - 1961 : Standard dimensions of paper covered rectangular copper conductor for transformer windings
- IS 2068 - 1962 : Standard dimensions of cotton covered rectangular copper conductor for transformer windings.
- IS 3454 - 1966 : Standard dimensions of paper covered round conductors used for transformer windings.
- IS 450 - 1964 : Standard dimensions of cotton covered round conductors for transformer windings.

1.10 GENERAL DESIGN PROCEDURE

- In general any electrical machine has two windings. The transformer has primary and secondary winding. The dc machine and synchronous machine has armature and field winding. The induction machine has stator and rotor winding.

- The basic principle of operation of all electrical machine is governed by Faraday's law of induction. Also in every electrical machine the energy is transferred through the magnetic field. Hence a general design procedure can be developed for the design of elelctrical machines.

- The general design procedure is to relate the main dimentions of the machine to its rated power output. An electrical machine is designed to deliver a certain amount of power called rated power. The rated power output of a machine is defined as the maximum power that can be delivered by machine safely. In dc machine the power rating is expressed in kW and in ac machine in kVA. In case of motor the output power is expressed in HP.

- In electrical machines the core and winding of the machine are together called active part. (Because the energy conversion takes place only in all active part of the machine).

- A general output equation can be developed for dc machine which relates the power output to volume of active part (D^2L), speed, magnetic and electric loading.

- Similarly a general output equation can be developed for ac machine which relates kVA rating to volume of active part (D^2L), speed, magnetic and electric loading.

1.11 BASIC PRINCIPLES

The action of electromagnetic machines can be related to three basic principles which are,

1. Induction
2. Interaction
3. Alignment

1. Faraday's Law of Electromagnetic Induction

This law states that emf induced in a closed electric circuit is equal to the rate of change of flux linkages.

Flux linkages $\psi = N\Phi$

N – the number of turns in a coil

Φ – flux linking with all of them

In most cases, the flux Φ does not link with all the turns or alternatively all the turns do not link with the same flux.

Total flux linkages

$$\Psi = N_1\Phi_1 + N_2\Phi_2 + N_3\Phi_3 \dots N_n \Phi_n$$

$$= \sum_{k=1}^{n} N_k \Phi_k \qquad \dots (1.1)$$

N_k = The number of turns which link with flux Φ_k

In this case there is a change in the value of the flux linkages of the coil,an induced emf is produced is given by,

$$e = -\frac{d\Psi}{dt} \text{ Volt} \qquad \dots (1.2)$$

(-) sign indicates that the direction of the induced emf

The Change in Flux Linkages can be Caused in Three Ways :

1. The coil is stationary with respect to flux and the flux varies in magnitude with respect to time.

2. The flux is constant with respect to time and is stationary and the coil moves through it.

3. Both the changes mentioned above occur together (ie)the coil moves through a time varying field.

Method 1 : Where the coil is stationary and the flux is time varying, an emf called transformer or pulsational emf is produced. No motion is involved. There is no energy conversion. The process that really takes place is energy transference. This principles used in transformers.

Method 2 : The flux cutting rule can be employed to illustrate the emf generated in a conductor moving in a constant stationary field. The emf generated in a conductor of length moving at right angles to a uniform, stationary, time invarying magnetic field.

$$e = Blv \text{ Volt}$$

$$B - \text{flux density ,wb/m}^2 \text{ (T)}$$

$$l = \text{length of conductor ,(m)}$$

$$v - \text{linear velocity of conductors (m/s)}$$

The generated emf in this case is called a "motioned emf". Emf generated due to motion of conductor ,since motion is involved in the production of this emf,the process involves electromechanical energy conversion.

This principle is utilize in rotating machines like DC, induction machines, synchronous machines.

Method 3 : A conductor or coil is moving across a stationary time varying magnetic field (flux)and therefore both transformer as well as motional emf are produced in the conductor or coil. This process involves both transformer and energy conversion.

This principle is utilized in the commutator machines.

2. Interaction Principle (Biot-Savart's Law)

This law gives the value of force produced on account of interaction between a magnetic field and a current carrying conductor

$$F = B\,l\,I \sin\alpha \text{ Newton} \qquad \ldots (1.3)$$

where,

B – Flux density, wb/m^2 (T)

l – length of conductor

i – current carried by conductor, A

α – Angle between the direction of current and the direction of magnetic field

The direction of force produced is perpendicular to both current and magnetic field. Conductor and magnetic field are perpendicular to each other, thus $\alpha = 90°, \sin 90° = 1$

$$F = B\,l\,I \text{ Newton}$$

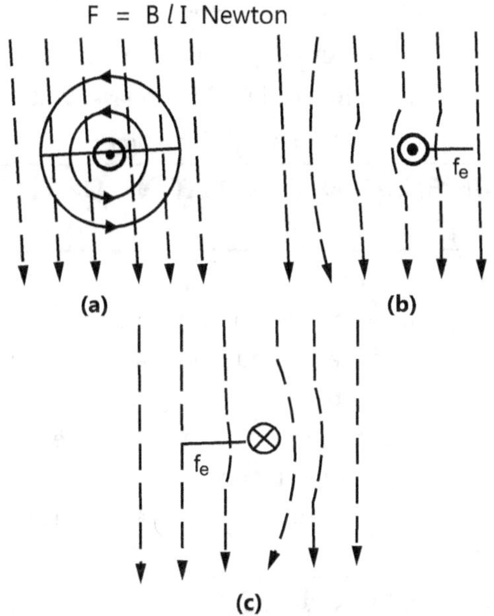

Fig. 1.2 : Force on a current carrying conductor situated perpendicular to a magnetic field (Interaction law)

- In Fig. 1.2 (a), B represents the flux density of an undisturbed (original) magnetic field. The introduction of a current carrying conductor introduces new magnetic field. The original field and the field due to conductor combine to produce a resultant field as shown in Fig. 1.2 (b). The resultant field is distorted in the neighbourhood of the conductor, the resultant flux density being greater on one side and lesser on the other and this results in production of an electromagnetic force in the direction indicated. In case the increase in flux density on one side is equal to the reduction on the other side, the electromagnetic force is given by Eq. 1.3.

- When either the direction of the current or the direction of the magnetic field is reversed. However, if the directions of both the current as well as the magnetic field are reversed, the direction of the force produced remains unaltered. Fig. 1.2 (c) shows the effect of reversing the current when the direction of the field is unchanged. It is clear that under these conditions the direction of force produced is reversed.

- Biot Savart's law can be applied to determine force between two current carrying conductors. Fig. 1.3 shows two parallel current carrying conductors of l separated by a distance D and situated in a medium of permeability μ. The two currents are I_1 and I_2. In Fig. 1.3 (a) the two currents flow in the same direction while in Fig. 1.3 (b) they flow in the opposite direction. The resultant magnetic fields are also shown. It is clear that when conductors carry currents in the same direction, there is a force of attraction between them, while there is a force of repulsion between them if they carry currents in the opposite directions.

The value of the flux density, at the position of conductor carrying current I_2,

Due to current I is :

Flux density $\beta = \mu H$

$$\mu = \frac{I_1}{2\pi D}$$

Electromagnetic force

$$F = B I l$$

$$F = \mu \frac{I_1}{2\pi D} I_2 l$$

$$F = \frac{\mu l}{2\pi D} I_1 I_2 \text{ Newton}$$

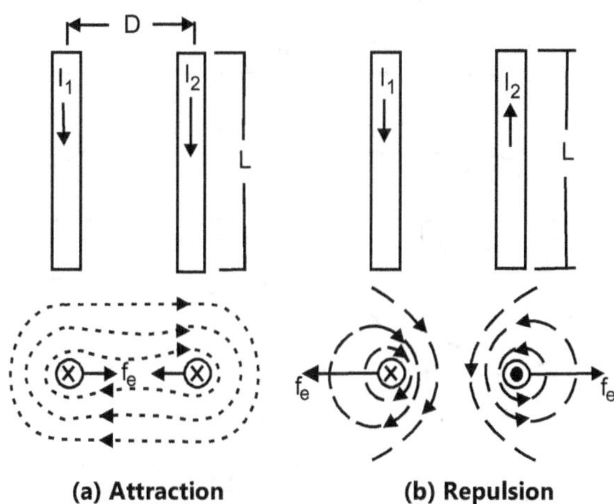

Fig. 1.3 : Forces between current carrying conductor

3. Alignment

If a magnetic field exists in a low permeability medium like air and if piece of high permeability material is placed in this field, the latter experiences a force which tries to align it with the direction of field in such away that it occupies a position of minimum reluctance. The principle of production of force due to alignment is used in reluctance motors.

1.12 MAIN DIMENSIONS OF ROTATING MACHINE

- In rotating machines the active part is cylindrical in shape. The volume of the cylinder is given by the product of area of cross section and length. If D is the diameter and L is the length of cylinder, then the volume is given by $\pi D^2 L/4$. Therefore D and L are specified as main dimensions.

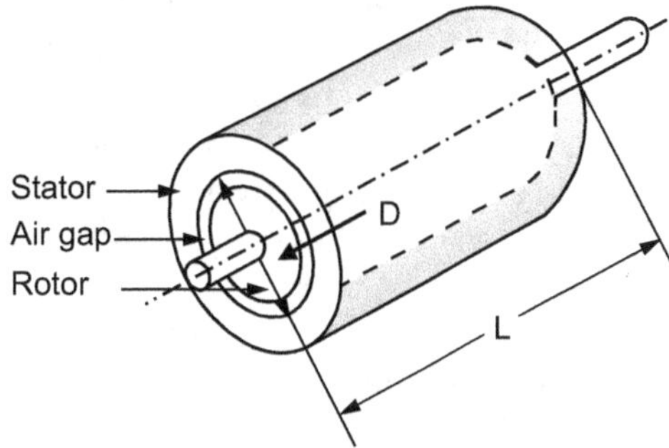

Fig. 1.4 : Main dimensions of rotating machines

- In case of dc machine, D represent the diameter of armature and L represent the length of armature. In case of ac machine, D represent the inner diameter of stator and L represent the length of stator core. The Fig. 1.5 shows the main dimensions of rotating machines

Here, D_r = Diameter of rotor

l_g = Length of air-gap

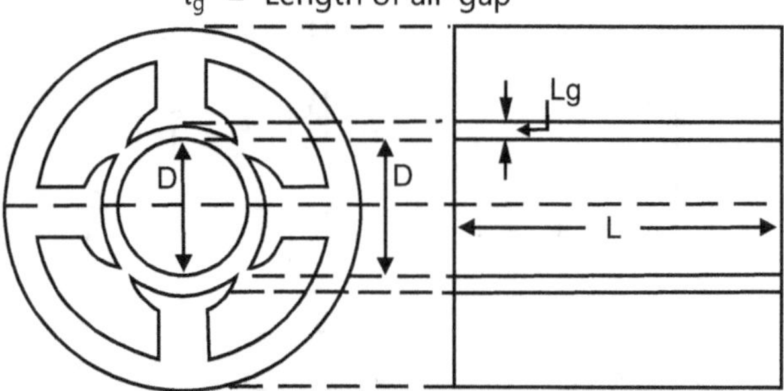

(a) Main dimensions of DC machines

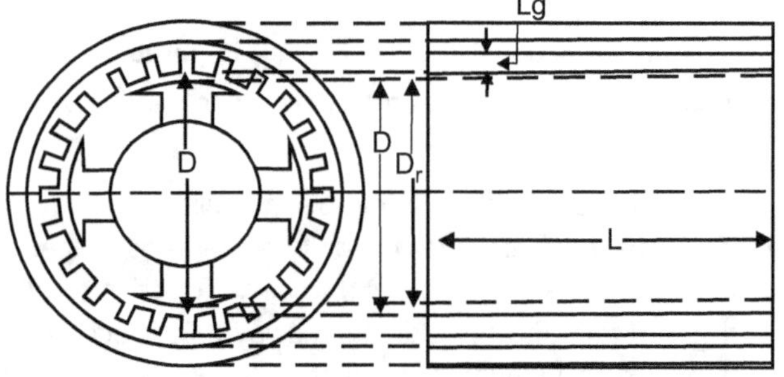

(b) Main dimensions of AC machines

Fig. 1.5 : Main dimensions of rotating machines

General Symbols Used for Designing of Induction Motor

D = Inner   diameter of stator or stator bore, m

L = Stator core length, m

n = Speed, rps;

n_s = synchronous speed, rps;

p = No. of poles;

Z = Total no. of armature or stator conductors

τ = Pole pitch, m;

T_{ph} = Turns per phase

I_z = Current in each conductor, Amp

K_w = Winding factor;

I_{ph} = Current per phase, Amp;

E_{ph} = Induced EMF per phase, Volt;

Q = kVA rating of machine

1.13 DIMENSIONS AND RATING OF THE MACHINE

The power rating of rotating machines is related to its main dimensions, namely the armature diameter and armature length. A few general equations are developed are applicable to all types of rotating machines like DC, induction and synchronous machines. However it must be mentioned that the design process of different machines cannot be demonstrated with a set of few general equations.

1.14 TOTAL LOADINGS

- **Total Magnetic Loading**

  The Total flux around the stator periphery at the air gap is called the total magnetic loading.

  Total magnetic load = $p\phi$

- **Total Electric Loading :**

  The total number of ampere conductors around the stator periphery is called the total electric loading.

  Total electric loading = $I_z Z$

- **Specific Magnetic Loading**

  The Average flux density over the air gap of a machine is known as specific magnetic loading and it is also defined as the ratio of total flux around the air gap and the area of flux path at the air gap.

$$B_{av} = \frac{\text{Total flux around the air gap}}{\text{Area of flux path at the air gap}} = \frac{p\phi}{\pi DL}$$

- **Specific Electric Loading**

It is defined as the ratio of total number of ampere conductors and the armature periphery at the air gap.

$$ac = \frac{\text{Total number of ampere conductors}}{\text{Armature periphery at the air gap}} = \frac{I_z Z}{\pi D}$$

The typical values of specific magnetic loading and electric loading for various types of rotating machines are listed in Table 1.1

Table 1.1

Machine	Specific Magnetic loading B_{av} in Wb/m²	Specific Electric Loading ac in amp.cond./m
DC machine	0.4 to 0.8	15000 to 50000
Induction motor	0.3 to 0.6	5000 to 45000
Synchronous machine	0.52 to 0.65	20000 to 40000
Turbo-alternator	0.52 to 0.65	50000 to 75000

1.15 OUTPUT EQUATION

"Expressed in terms of its main dimensions, specific loadings and speed."

- **Direct Current Machines**

Power level developed by armature in kW

$$P_a = \text{generated EMF} \times \text{armature current} \times 10^{-3}$$
$$= E.I_a \times 10^{-3}$$

But,
$$E = \phi \cdot Z \cdot n \frac{p}{a}$$

$\therefore$
$$P_a = \phi \cdot Z \cdot n \frac{p}{a} I_a \times 10^{-3}$$
$$= (p \cdot \phi)\left(\frac{I_a}{a} \cdot Z\right) n \times 10^{-3} \quad \left(\text{since } I_z = \frac{I_a}{a}\right)$$
$$= (P \cdot \phi)(I_z \cdot Z) n \times 10^{-3}$$

Hence,
$$P_a = (\text{total magnetic loading}) (\text{total electric loading}) (\text{speed in rps}) \times 10^{-3}$$

Specific magnetic loading $B_{av} = \dfrac{P \cdot \phi}{\pi \cdot D \cdot L}$ or

$$p \cdot \phi = \pi \cdot D \cdot L \cdot B_{av}$$

By substituting these values we get,

Specific electric loading $ac = \dfrac{I_z \cdot Z}{\pi \cdot D}$ or $I_z \cdot Z = \pi \cdot D \cdot ac$

$$P_a = (\pi \cdot D \cdot L \cdot B_{av}) (\pi \cdot D \cdot ac) n \times 10^{-3}$$
$$= C_o D^2 Ln$$

where, $C_o = \pi^2 \cdot B_{av} \cdot ac \times 10^{-3}$

C_o = Output co-efficient

- **Alternating Current Machines**

Output Equation:"Expressed in terms of its main dimensions, specific loadings and speed."

Consider,   m = number of phases and having 1 circuit (parallel path) per phase

kVA rating of machine:

Q = no. of phases × o/p voltage / phase × current / phase × 10^{-3}

$$= mE_{ph}I_{ph} \times 10^{-3}$$

Terminal voltage/phase may be taken as equal to induced EMF/phase.

We have,

$$E_{ph} = 4.44 f\phi T_{ph}K_w.$$

$\therefore$
$$Q = m \times 4.44 f\phi T_{ph}K_w \times I_{ph} \times 10^{-3}$$

But, $f = p.n_s/2$

$\therefore$
$$Q = m \times 4.44 \left(\frac{p \cdot n_s}{2}\right) \phi T_{ph}K_w I_{ph} \times 10^{-3}$$
$$= 1.11 k_w (p\phi \; 2m \; I_{ph} \; T_{ph}) n_s \times 10^{-3}$$

Current in each conductor $I_z = I_{ph}$ (as only one circuit/phase)

Total no. of armature conductors

Z = no. of phases × (2 × turns per phase)
$$= 2mT_{ph}$$

$\therefore$ Total electric loading $= I_z Z = 2mI_{ph}T_{ph}$

Hence, $Q = 1.11 K_w(p\phi)(I_z Z)n_s \times 10^{-3}$
$$= 1.11K_w(\text{total magnetic loading}) (\text{total electric loading}) (\text{synchronous speed}) \times 10^{-3}$$

But, $p\phi = \pi DL \cdot B_{av}$ and $I_z Z = \pi D \cdot ac$

$\therefore$
$$Q = 1.11 K_w (\pi DL \cdot B_{av}) (\pi D \cdot ac)n_s \times 10^{-3}$$
$$= (1.11\pi^2 B_{av}acK_w \times 10^{-3})D^2 Ln_s$$
$$= (11 B_{av}acK_w \times 10^{-3})D^2 Ln_s$$
$$= C_o D^2 Ln_s$$

where, $C_o = 11B_{av}acK_w \times 10^{-3}$
$$= \text{Output co-efficient}$$

The capacity of motor is usually given either in horse power (h.p) or in Kw.But this has to be changed in kVA

Input kVa, $Q = \dfrac{kW}{\eta \cdot \cos\phi}$

The rating of induction motor is given in Horse Power

$\therefore$
$$Q = \frac{HP \times 0.746}{\eta \cdot \cos\phi}$$

Full load p.f. usually lies 0.82 to 0.92

High p.f. is generally obtained with high speed motor.

The full load efficiency is usually varies between 0.82 to 0.93

- **Factors Affecting Size of a Rotating Machine:**

Speed:

According to the output equation of a rotating machine with a given value of kVA rating

$$Q = C_o D^2 L n_s$$

$$D^2 L = \frac{Q}{C_o n_s} \Rightarrow \text{Represent the volume of machine}$$

$$\text{Speed,} \quad n_s \propto \frac{1}{D^2 L}$$

It is clear from output equation that the volume of active parts is inversely proportional to the speed. Increase in speed means less volume (smaller size) that is low cost.

Output Co-Efficient:

According to the output equation of a rotating machine with a given value of speed

$$Q = C_o D^2 L n_s$$

Output Co-efficient,

$$C_o \propto \frac{1}{D^2 L}$$

It is clear from output equation that the volume of active parts is inversely proportional to the value of output coefficient C_o. speed. So Increase in value of c_o results in reduction in size and cost of the machine..

$$C_o = \text{Output co-efficient}$$

1.16 CHOICE OF SPECIFIC LOADINGS

1.16.1 Choice of Specific Magnetic Loadings

The choice if specific magnetic loading is influence by certain factors. Some are general in nature and apply to all types of machines and some are specific and apply to individual machines.

Some general factors are:

1. Maximum flux density in iron parts of machine

2. Magnetizing current

3. Core losses

1. Maximum Flux Density in Iron

- The maximum flux density in any iron part of machine must be below a certain limiting value. The maximum flux density occurs in the teeth of the armature (or stator core). [Teeth are the portion of the core in between slots].

- The flux density in the teeth is directly proportional to specific magnetic loading. Hence the choice of specific loading should be such that the maximum value of flux density in the teeth is not exceeded. The maximum value of flux density in the teeth is not exceeded. The maximum value of flux density in the teeth is between 1.7 to 2.2 Wb/m^2.

2. Magnetizing Current

- The magnetizing current of a machine is directly proportional to mmf. The mmf is directly proportional to specific magnetic loading. Hence a large value of specific magnetic loading results in increased values of magnetizing mmf and magnetizing current.

- The value of magnetizing current is not usually a serious design consideration in dc machines. But in induction motors an increased value of magnetizing current results in low power factor. Hence specific magnetic loading in induction motors is lower than in dc machines. For synchronous machiens the magnetizing current is not so critical and the value of specific magnetic loading is intermediate between that of dc and induction machines.

3. Core Loss

- The core loss in any part of the magnetic circuit is directly proportional to flux density for which it is going to be designed. The flux density is directly proportional to the specific magnetic loading. Hence the core loss in a machine varies directly as the specific magnetic loading. Thus a large value of specific magnetic loading results in increased core loss and consequently a decreased efficiency and an increased temperature rise.

- With a given specific magnetic loading, the core loss increases as the frequency of flux reversals is increased. This is because the hystersis loss is directly proportional to the frequency and eddy current loss is proportional to the square of the frequency. It follows that for high speed dc machines, or high frequency ac machines, specific magnetic loadings must be reduced in order to achieve lower iron loss.

1.16.2 Choice of Specific Electric Loading

1. **Copper Loss and Temp Rise:** large value of ac , needs greater amount of copper, results in higher copper losses and large temperature rise

2. **Voltage:** for high voltage machines, less value of ac should be chosen, because it needs large space for insulation.

Let, w_s = Width of the slot

d_s = Depth of the slot

S_f = Slot space factor

y_s = Slot pitch

δ = Current density

The specific electric loading can be related to the above terms by the equation,

$$ac = d_s\,(w_s/y_s)\,\delta\,S_f \qquad \text{... (1.4)}$$

From equation (1.4) it is clear that the specific loading is directly proportional to slot space factor S_f. In high voltage machines, greater insulation thickness is required and therefore the space factor for these machines is lower. Hence an increase in voltage will, in general, necessitate a reduction in specific electric loading ac.

3. **Overload Capacity:** larger value of ac , results in large number of turns per phase. Which in turn increase the leakage reactance of the machine, reduces the overload capacity of the machine.

4. **Permissible Temperature Rise**

Let, θ = Temperature rise

S = Dissipating surface

Q = Loss dissipated

c = Cooling Coefficient

δ = Current density

ρ = Resistivity

$$\left.\begin{array}{l}\text{Loss dissipated per}\\ \text{unit area of}\\ \text{armature surface}\end{array}\right\}\; q = \frac{(\text{current})^2 \times \text{No. of conductors} \times \text{Resistance}}{\text{Surface area of armature}}$$

$$= \frac{I_z^2 \times Z \times \rho\, L\,/a_z}{\pi D L}$$

$$= \frac{I_z Z}{\pi D} \times \frac{I_z}{a_z} \times \rho = ac\,\delta\rho \qquad \text{... (1.5)}$$

where, $\delta = I_z/a_z$

Also, $q = Q_l/S$

The temperature rise, $\theta = \dfrac{Q_l\,c}{S} = qc$

$$q = \theta/c$$

$$ac\,\delta\rho = \frac{\theta}{c}$$

$\therefore$ Maximum allowable specific electric loading,

$$ac = \frac{\theta}{\rho\,\delta\,c} \qquad \text{... (1.6)}$$

It can be inferred that the heat dissipated per unit area of armature is proportional to specific electric loading.

From equation (1.6) it is clear that allowable specific electric loading is fixed by allowable temperature rise and the cooling coefficient. A high value of ac can be used in a machine when a high temperature rise is allowed. The maximum allowable temperature rise if a machine is determined by the type of insulating materials used in it. When better quality insulating materials which can withstand high temperature rises are used in the machines, increased values of specific electric loading can be used. This results in reduction in the size of the machine.

A high value of electric loading may be used if the cooling coefficient of the machine is small. The value of cooling coefficient depends upon the ventilation conditions in the machine. High speed machines will have better ventilation and so higher value of ac can be used.

5. **Size of Machine**

From the equation (1.6) it is clear that ac depends on the dimension of the slot. For large machines the depth of the slot will be greater and so higher values of ac can be used. Actually if the current density and the slot space factor are assumed constant, then specific electric loading is proportional to the diameter as slot depth usually depends upon the diameter.

6. **Current Density**

From the equation, $q = ac\,\delta\,\rho$ it is clear that a higher value of specific electric loading can be used in a machine which employs lower current density in its conductors. (because $ac = q/\delta\,\rho$).

Typical values of current density are in the range of 2 to 5 A/mm^2. The temperature rise is usually 40°C (above ambient) for normal applications and cooling coefficient is between 0.02 and 0.035°C W-m^2.

SOLVED EXAMPLES

Example 1.1 : *A 350 kW, 500 V, 450 rpm, 6-pole, dc generator is built with an armature diameter of 0.87 m and core length of 0.32 m. The lap wound armature has 660 conductors. Calculate the specific electric and magnetic loadings.*

Solution : Given :

$$P = 350 \text{ kW}$$
$$n = 450/60 \text{ rps}$$
$$D = 0.87 \text{ m}$$
$$Z = 660$$
$$V = 500 \text{ V}$$
$$p = 6$$
$$L = 0.32 \text{ m}$$

Lap wound

Specific electric loading,

$$ac = \frac{I_z Z}{\pi D}$$

Specific magnetic loading,

$$B_{av} = \frac{p\phi}{\pi DL}$$

The power output of the generator,

$$P = VI \times 10^{-3} \text{ in kW}$$

$\therefore$ full load current,

$$I = \frac{P}{V \times 10^{-3}} = \frac{350}{500 \times 10^{-3}} = 700 \text{ amps}$$

Neglecting field current,

$$I_a \approx I$$

$$\left.\begin{array}{l}\text{current through}\\ \text{each armature}\\ \text{conductor}\end{array}\right\} I_z = \frac{\text{armature current}}{\text{No. of parallel paths}} = \frac{I_a}{a}$$

$$= \frac{700}{6} = 116.67 \text{ amps}$$

(a = p in lap wound)

specific electric loading,

$$ac = \frac{I_z Z}{\pi D} = \frac{116.67 \times 660}{\pi \times 0.87}$$

$$= 28173 \text{ amp.cond./m}$$
$$ac = 28173 \text{ amp.cond./m}$$

Induced emf in dc generator,

$$E = \phi Z n \frac{p}{a} = \phi Z n, \text{ for lap winding}$$

$$(\because p = a)$$

Hence, flux per pole,

$$\phi = \frac{E}{Zn} \approx \frac{V}{Zn} = \frac{500}{660 \times 450/60}$$

$$= 0.101 \text{ Wb}$$

Specific magnetic loading,

$$B_{av} = \frac{p\phi}{\pi DL} = \frac{6 \times 0.101}{\pi \times 0.87 \times 0.32}$$

$$= 0.6929 \text{ Wb/m}^2$$

1.17 ELECTRICAL ENGINEERING MATERIALS

Materials used in Electrical machines are classified into three types:

1. Conducting;
2. Magnetic
3. Insulating

Design of electrical machines depends mainly on quality of materials used. If low quality materials are used, the machine will be less efficient, more bulky, higher weight and higher cost. Operational running cost will also be higher. A designer should have perfect knowledge of properties and cost of these materials so that the design can be both efficient and cost-effective.

1.17.1 Conducting Materials

Conducting Materials are of Following Categories

1. High conductivity materials
2. High resistivity materials
3. Electrical carbon materials
4. Super conducting materials

1. **High Conductivity Materials (Low Resistance):** Used for windings of electrical machines and equipments. Material with lowest resistance should be selected so that it contributes lowest ohmic losses to enhance efficiency and to reduce temp-rise.

Requirements of high conductive materials:

- Highest Possible conductivity (least Resistance)
- Least possible temperature coefficient of resistance
- Adequate resistance to corrosion
- Adequate mechanical strength and high tensile strength
- Suitable for jointing by brazing/soldering/welding so that the joints are highly reliable contributing lowest resistance.
- Suitable for roll ability, draw ability, so that conductors of required shape (wire/strip) are easily manufactured.

Material: Copper, Aluminium, Iron steel, Alloys of copper

Copper : Relative immunity from oxidation and corrosion, highly malleable & ductile metal - it can be cast, forged, rolled, drawn, machined, easily soldered. Annealed high conductivity copper and hard drawn copper wires are used for windings of electrical machines

Aluminium : The existing copper deposits are fast exhausting and the price of copper fluctuates widely, therefore it replaces copper in many applications. Aluminium when used in small transformer, decreases overall cost (comparatively lesser than copper) and weight (approximately 3.3 times lighter than copper) Aluminium when used in large transformer, increases overall cost (its resistivity 1.62 times higher than copper and size (its volume approximately 2.04 times greater than copper) (while designing due account should be taken of their differences in resitivity, cost, weight, conductivity, temp. etc.,)

Iron and Steel : Steel alloyed with chromium and Aluminium (robustness with good heat dissipation) – for making starter rheostats. Cast iron – resistance grids to be used in starters of large motors.

Alloys of Copper : Bronze (copper base alloys containing tin, cadmium & berryllium) possesses higher resistivity and mechanical strength. Beryllium copper – current carrying springs, brush holders, sliding contacts and knife switch blades. Cadmium copper- contact wires and commutator segments. Brass (66% Cu and 34% Zinc) has greater mechanical strength ,wear resistance and lower conductivity – widely used in current carrying materials . * Copper silver alloy (99.10% and 0.06% to 0.1%) has resistance to thermal shortening and creep (turbo-alternators)

Best conducing material is silver. Next best is copper and then aluminum. Properties of these are compared in the following table.

Table 1.2

Sr. No.	Property	Unit	Silver	Copper	Aluminium
1	Conductivity	--	1.0	0.975	0.585
2	Resistivity	$\mu\Omega$-cm	1.46	1.777	2.826
3	Temp-coeff.	% per °C	0.337	0.393	0.4
4	Cost	---	Prohibitively high	Medium	Low

2. **Material of Low Conductivity (High Resistivity):** It is usually called high resistance conductors as resistors, resistance coils, resistance elements or heating elements- are used to dissipate electrical energy as heat i.e., in starting and regulating devices for motors. Used for heating devices, thermo couples, resistance etc

Categories According to their Purposes

I Group : Materials used for Precision work- for making standard resistances and resistance boxes.

Properties

- Stability of resistance over the period of time and during fluctuations of temperature.
- Low temperature co-efficient.
- Minimum thermo electric effect at contact of material -does not introduce errors into measurements.

Material : Manganin (Cu 86%, Mn 12%and Ni 2%)

II Group : Materials used for making rheostats

Properties

- Should have large thermo-emf
- Large resistance temperature co-efficient.

Special Requirement :

- High permissible working temperature and low cost

Material : Constantan (Cu 65% and 40 to 35% Ni) Sometimes small amounts of manganese and iron also included.

III Group : Materials used for making heating devices in electric furnaces and loading rheostats

Special Requirements

- High permissible working temperature, low cost and should have non-corrosive

Material : Nichrome (Nickel, chromium and iron) (optimum working temperature-900°-to 1000°)

3. **Electrical Carbon Materials :** Electrical carbon materials are made from graphite and other forms of carbon coal.

Properties

- Negative temp. Co-efficient (contact voltage drop decreases with increasing temp)
- Low wear and tear (due to self lubricant property)

Material : Carbon, carbon graphite, graphite, electro graphite, metal graphite- used for making brushes for electrical machines.

4. **Super Conducting Materials :** Materials exhibiting zero value of resistivity are known as Super conductors. A large number of metals become super-conducting below a particular temperature characteristic of the particular metal. This temperature is known as the transition temperature.

For Example: Alluminium – Trans.temp.- 1.18 K Uranium – 0.80 K It is interesting to note that copper, silver, gold etc., are very good conductors at room temp., but do not exhibit superconducting properties. (vice versa for other metals and alloys) Application of superconductor It can be used for the transformers and rotating electrical machines, depending upon the comparative gain in the reduction of full load copper losses against the cost for provision of cryogenic conditions

1.17.2 Magnetic Materials

- All magnetic materials possess magnetic properties to a greater or a lesser degree. The magnetic properties of materials are characterized by their relative permeability. In accordance with the value of relative permeability, materials may be divided into three broad classes.

  1. **Ferromagnetic Materials :** The relative permeabilities of these materials are much greater than unity and these permeability values are dependent upon the magnetizing force
     Relative permeability $-\mu_r \gg 1$ (Nickel, cobalt, iron, steel, silicon steel etc.,)

  2. **Paramagnetic Materials :** These materials have their relative permeabilities only slightly greater than unity. The value of susceptibility, is thus positive for these materials.
     Relative permeability $-\mu_r > 1$ (Air, Alluminium, palladium etc.,)

  3. **Diamagnetic Materials :** These materials have their relative permeabilities slightly less than unity. In both Paramagnetic and Diamagnetic materials the value of permeability is independent of the magnetizing force.
     Relative permeability $-\mu_r < 1$ (Bismuth, silver, lead, copper, water etc.,)

Ferromagnetic materials are very useful for electrical engineering applications. Why?

When a ferromagnetic material is placed in a magnetic field, there is considerable distortion and, therefore, the force exerted is very large. This property makes ferro-magnetic materials very useful for electrical engineering applications.

Example: Iron, nickel, cobalt, and many of their alloys are ferromagnetic.

Types of Magnetic Materials

1. Soft Magnetic Material
2. Hard Magnetic Materials

1. Soft Magnetic Material

- The hysteresis loss depends upon the area of hysteresis loop. For this reason, magnetic cores used in alternating magnetic fields are made from materials whose hysteresis loops are more or less narrow [see Figs. 1.6 (a) and (b)]. (Silicon steel, nickel- iron alloys etc.)

These materials are called soft magnetic materials. Soft magnetic materials are used in the manufacture of electrical machines, transformers and many kinds of electrical apparatus, instruments and devices.

- **Classifications:**
  - ➤ Solid core materials
  - ➤ Laminated core material for pulsating fluxes
  - ➤ Electrical sheet and strip
  - ➤ Special purpose alloys

- **Solid Core Materials :** For steady fluxes. These materials are normally used for parts of magnetic circuits carrying steady flux as cores of d.c . Electromagnets, relays and field frame (i.e., yoke) of d.c. machines. Examples: cast iron, cast steel and Ferro-cobalt

- **Laminated Core Material for Pulsating Fluxes :** These materials are normally used for parts of magnetic circuits carrying pulsating flux as cores of d.c. armature, stator and rotor of ac machines

- **Electrical Steel Sheets Dynamo Grade Steels :** Low silicon content are used in rotating electrical machines

- **Transformer Grade Steels :** High silicon content are used in transformers. **Cold rolled grain oriented steel (CRGO)** - is suitable for use in large transformers and turbo alternators

- **Special Purpose Alloys :** Are used in instrument transformers, induction coils and choke. Example : permalloys, superpermalloy, perminvar etc.,

2. Hard Magnetic Materials :

- Materials with broad hysteresis loops [Fig 1.6 (c)] are called hard magnetic materials. These materials are used in certain types of electrical machines. rating, and in all kinds of instruments and devices requiring permanent magnets which set up magnetic fields of their own.

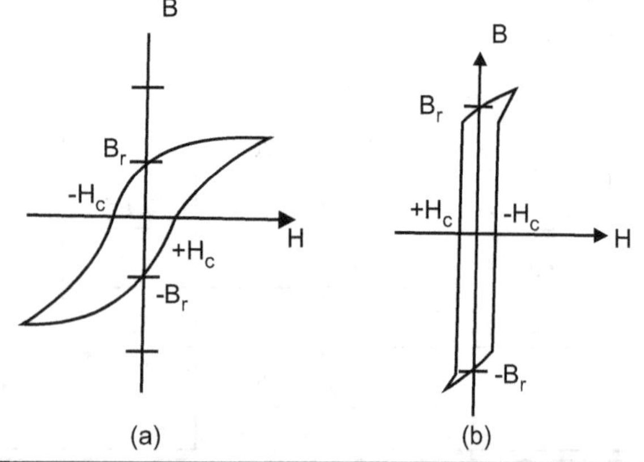

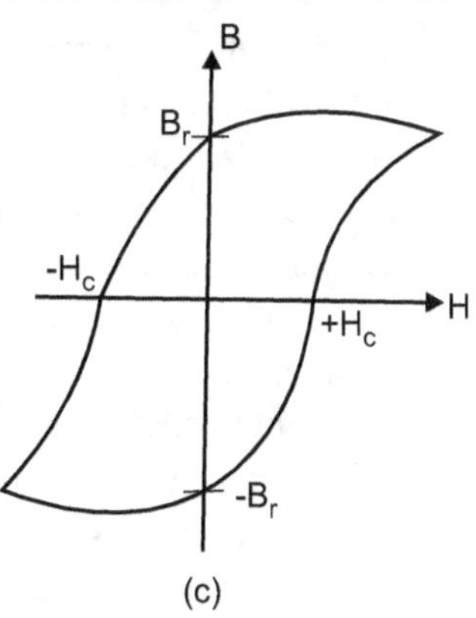

Fig. 1.6 : Hysteresis loops

- Hard or permanent magnetic materials have large size hysteresis loop (obviously hysteresis loss is more) and gradually rising magnetization curve.

  Ex.: carbon steel, tungsten steal, cobalt steel, alnico, hard ferrite etc.

- **Properties of Good Magnetic Material :**

  The some of the properties that a good magnetic material should possess are listed below.

  - Low reluctance or should be highly permeable or should have a high value of relative permeability μ_r.
  - High saturation induction (to minimize weight and volume of iron parts)
  - High electrical resistivity so that the eddy emf and the hence eddy current loss is less
  - Narrow hysteresis loop or low coercivity so that hysteresis loss is less and efficiency of operation is high
  - A high curie point. (Above Curie point or temperature the material loses the magnetic property or becomes paramagnetic, that is effectively non-magnetic)
  - Should have a high value of energy product (expressed in joules / m^3).

1.18 INSULATING MATERIALS

To avoid any electrical activity between parts at different potentials, insulation is used. An ideal insulating material should possess the following properties.

- Should have high dielectric strength.
- Should with stand high temperature.
- Should have good thermal conductivity
- Should not undergo thermal oxidation
- Should not deteriorate due to higher temperature and repeated heat cycle
- Should have high value of resistivity (like 1018 Wcm)

- Should not consume any power or should have a low dielectric loss angle d
- Should withstand stresses due to centrifugal forces (as in rotating machines), electro dynamic or mechanical forces (as in transformers)
- Should withstand vibration, abrasion, bending
- Should not absorb moisture
- Should be flexible and cheap

Factors Affecting the Electrical Properties

- Dimensions of test piece
- r.m.s. value, wave form and frequency of impressed voltage.
- Temperature and moisture content of test piece
- Mechanical pressure on test piece

Applications of Insulating Materials Employed for the insulation of

- Wires for magnet coils and windings of machine
- Laminations for Machines and transformers

Insulating Material Classification

- Insulating materials can be classified as
  - Solid
  - Liquid
  - Gas
  - Vacuum.
- The term insulting material is sometimes used in a broader sense to designate also insulating liquids, gas and vacuum.

Solid: Used with field, armature, transformer windings etc. The examples are:

- Fibrous or inorganic animal or plant origin, natural or synthetic paper, wood, card board, cotton, jute, silk etc., rayon, nylon, terelane, asbestos, fiber glass etc.,
- Plastic or resins. Natural resins-lac, amber, shellac etc., Synthetic resins-phenol formaldehyde, melamine, polyesters, epoxy, silicon resins, bakelite, Teflon, PVC etc
- Rubber : natural rubber, synthetic rubber-butadiene, silicone rubber, hypalon, etc.,
- Mineral : mica, marble, slate, talc chloride etc.,
- Ceramic : porcelain, steatite, alumina etc.,
- Glass : soda lime glass, silica glass, lead glass, borosilicate glass
- Non-resinous : mineral waxes, asphalt, bitumen, chlorinated naphthalene, enamel etc.,

Liquid: Used in transformers, circuit breakers, reactors, rheostats, cables, capacitors etc., & for impregnation. The examples are:

- Mineral oil (petroleum by product)
- Synthetic oil askarels, pyranols etc.,
- Varnish, French polish, lacquer epoxy resin etc.,

Gaseous: The examples are:
- Air used in switches, air condensers, transmission and distribution lines etc.,
- Nitrogen use in capacitors, HV gas pressure cables etc.,
- Hydrogen though not used as a dielectric, generally used as a coolant
- Inert gases neon, argon, mercury and sodium vapors generally used for neon sign lamps.
- Halogens like fluorine, used under high pressure in cables

  No insulating material in practice satisfies all the desirable properties. Therefore a material which satisfies most of the desirable properties must be selected.

Classification of Insulating Materials Based on Thermal Consideration

- The insulation system (also called insulation class) for wires used in generators, motors transformers and other wire-wound electrical components is divided into different classes according the temperature that they can safely withstand.
- As per Indian Standard (Thermal evaluation and classification of Electrical Insulation, IS. No. 1271, 1985, first revision) and other international standard insulation is classified by letter grades A, E, B, F, H (previous Y, A, E, B, F, H, C).

Table 1.3

Sr. No.	Insulation Class	Maximum Operating Temperature in °C	Typical Materials
1	Y	90	Cotton, silk, paper, wood, cellulose, fiber etc., without impregnation or immersed
2	A	105	The material of class Y impregnated with natural resins, cellulose esters, insulating oils etc., and also laminated wood, varnished paper etc.
3	E	120	Synthetic resin enamels of vinyl acetate or nylon tapes, cotton and paper laminates with formaldehyde bonding etc.
4	B	130	Mica, glass fiber, asbestos etc., with suitable bonding substances, built up mica, glass fiber and asbestos laminates.
5	F	155	The materials of Class B with more thermal resistance bonding materials.
6	H	180	Glass fiber and asbestos materials and built up mica with appropriate silicone resins.
7	C	> 180	Mica, ceramics, glass, quartz and asbestos with binders or resins of super thermal stability.

- The maximum operating temperature is the temperature the insulation can reach during operation and is the sum of standardized ambient temperature i.e. 40 degree centigrade, permissible temperature rise and allowance tolerance for hot spot in winding. For example, the maximum temperature of class B insulation is (ambient temperature 40 + allowable temperature rise 80 + hot spot tolerance 10) = 130°C.
- Insulation is the weakest element against heat and is a critical factor in deciding the life of electrical equipment. The maximum operating temperatures prescribed for different class of insulation are for a healthy lifetime of 20,000 hours. The height temperature permitted for the machine parts is usually about 2000C at the maximum. Exceeding the maximum operating temperature will affect the life of the insulation. As a rule of thumb, the lifetime of the winding insulation will be reduced by half for every 10°C rise in temperature. The present day trend is to design the machine using class F insulation for class B temperature rise.

EXERCISE

1. Name the basic structural parts of an electromagnetic rotating machine.
2. What are the factors those limit the design of a machine?
3. Classify electrical machines from the view point of manufacturing process.
4. How the Faraday's law of electromagnetic induction principle is applied in design of an electrical machine?
5. Briefly explain Biot-Savart's law as applicable in electric machine design.
6. What are the main dimensions in machine design ?
7. What do you mean by specific loading ?
8. Develop the output equation of
   (i) d.c. machine
   (ii) a.c. machine
9. What is output coefficient of machine ?
10. What are the factors those affect the size of rotating machines?
11. What are fundamental requirements of high conductivity material?
12. Write short note on magnetic material.

DESIGN OF ELECTRICAL APPARATUS

2.1 INTRODUCTION

In this chapter, details are given how to design the heating elements (coils), different types of starters for ac and dc motors, field regulators, electrical devices like field coils, chokes and electromagnets (lifting magnets).

2.2 DESIGN OF HEATING COILS

2.2.1 Heating Elements

This topic deals with design of heating elements used for commercial purposes. The material used for making heating elements is usually Nichrome. Mostly the configuration of heating elements are either round wire type or ribbon type (flat) elements. The following deals with the analytical part of design of heating elements.

2.2.2 Design of Round Wire Elements

In high temperature furnaces, the heat produced in the elements is mainly dissipated by radiation as the temperatures are very high.

Let P = electrical input, W ;

V = voltage applied, V;

R = resistance of element, Ω;

ρ = resistivity of element, Ω/m and mm^2 ;

d = diameter of wire, mm;

l = length of element, m.

Heat input

$$P = \frac{V^2}{R} = \frac{\pi}{4} \cdot \frac{V^2 d^2}{\rho L}$$

Dissipating surface of wire

$$= \pi \, dl \times 10^{-3} \ m^2$$

Heat input of wire surface

$$= \frac{\pi}{4} \cdot \frac{V^2 d^2}{\rho L} \cdot \frac{1}{\pi dl \times 10^{-3}}$$

$$= \frac{V^2 d \times 10^3}{4\rho l^2} \ W/m^2 \qquad \dots (2.1)$$

Now, q_{rad} = heat radiated per unit surface of wire

$$= 5.7 \times 10^{-8} \, e\eta \, (T_1^4 - T_2^4) \ W/m^2 \qquad \dots (2.2)$$

Here the term η has been included to take into consideration the effective value of emissivity.

We have η = radiating efficiency

= 1 for single elements and the value may go down to 0.5 for multiple element units

and e = 0.9 for heating elements

We have,

$$\frac{d}{l^2} = \frac{4 \, \rho q_{rad}}{V^2} \times 10^{-3} \qquad \dots (2.3)$$

but, $$\frac{V^2}{P} = R = \frac{4\rho L}{\pi d^2}$$

or $$\frac{l}{d^2} = \frac{\pi}{4} \cdot \frac{V^2}{\rho P} \qquad \dots (2.4)$$

Using eq. (2.3) and (2.4) the length and diameter of wire can be calculated.

| SOLVED EXAMPLES |

Example 2.1 : *A 1 kW, 250 V single element electric furnace is to employ a nicrome resistance wire operating at 1000 ℃. Estimate a suitable diameter and length of wire. Assume radiating efficiency = 1, emmisivity = 0.9 and and resistivity of wire 0.424 Ω mt at 1000 ℃, The ambient temperature is 20 ℃.*

Solution : $q_{rad} = 5.7 \times 10^{-8} \, e\eta \, (T_1^4 - T_2^4)$

$$= 5.7 \times 10^{-8} \times 0.9$$
$$\times 1 \, [(1000 + 273)^4 - (20 + 273)^4]$$
$$= 135 \times 10^3 \ W/m^2$$

As, known total heat

$$= q_{rad} \, (\pi dl \times 10^{-3})$$
$$= 135 \times 10^3 \, (\pi \times dl \times 10^{-3})$$
$$= 424 \, dl$$
$$= 1000 \ (given)$$

or $dl = 2.31$

Resistance $R = V^2/P = (250)^2/1000 = 62.5 \ \Omega \qquad \dots (1)$

But $$R = \frac{\rho l}{(\pi/4)d^2} = \frac{0.424 \, l}{(\pi/4)d^2} = 0.54 \frac{l}{d^2}$$

$$= 62.5 \ \Omega \ (calculated \ above)$$

$\therefore \quad l/d^2 = 115 \qquad \dots (2)$

(solving eq. (1) and (2) we have
$d = 0.272$ mm and $l = 8.5$ m)

2.3 STARTERS AND REGULATORS

2.3.1 Motor Starter

- The function of motor starters is to prevent the flow of excessive currents at starting and thus prevent undue large mechanical stresses which would otherwise act on machine parts. At the same time, the starter should allow a current high enough to produce a good starting torque so that the motor starts up against the specified load.

- The starter may take up the form of a liquid rheostat whose resistance can be gradually varied or it may be a metallic resistance starter where the resistance is varied in steps. In the case of starters with resistance steps, the current is taken to fluctuate between fixed upper and lower current limits. The method of calculation of resistance steps is given below.

2.3.2 Calculation of Resistance Steps

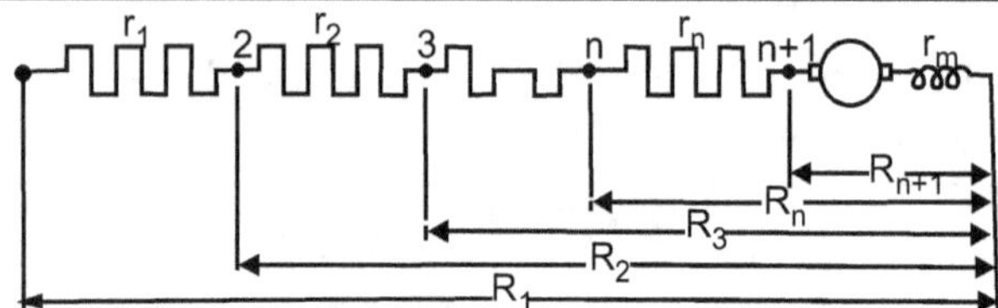

Fig. 2.1 : Resistance steps of starter

Referring to Fig. 2.1

Consider　I_1, I_2 = upper and lower limits of current respectively

$$\alpha = \frac{I_2}{I_1} \qquad \text{... (2.5)}$$

ϕ_1, ϕ_2 = useful flux per pole due to currents I_1 and I_2

$$\beta = \frac{\phi_1}{\phi_2} \qquad \text{... (2.6)}$$

$$\delta = \alpha\beta = \frac{\phi_1 I_2}{\phi_2 I_1} \qquad \text{... (2.7)}$$

n = number of resistance sections or elements

$n + 1$ = number of studs

$R_1, R_2, R_3 \dots R_{n+1}$ = total resistance in circuit, on 1st, 2nd, 3rd … $(n + 1)^{th}$ stud

$r_1, r_2, r_3 \dots$ = $(R_1 - R_2), (R_2 - R_3), (R_3 - R_4) \dots$

　　　= resistance of sections or elements

r_m = motor resistance

V = motor or line voltage, E = motor emf

s = slip, and X = reactance.

1.　Starters for DC Shunt Motors :

Shunt motors are constant flux machines $\phi_1 = \phi_2$ and $\beta = 1$.

To start with, the contact arm is put at stud 1. Hence limit the starting current to I_1, the resistance the circuit is,

$$R_1 = \frac{V}{I_1} \qquad \text{... (2.8)}$$

When, the motor starts rotating and builds up back emf and the current drops to its minimum value I_2. Assuming the back emf at stud 1 being E_1,

$$\therefore \quad I_2 = \frac{V - E_1}{R_1} \qquad \text{... (2.9)}$$

When the current has reached a value I_2, the resistance r_1 of the first section is cut out and the contact arm is moved to stud 2. As the resistance is cut out, the current rises to its maximum value, I_1. The back emf at stud 2 be E_2.

$$\therefore \quad I_1 = \frac{V - E_2}{R_2} \qquad \text{... (2.10)}$$

The back emf is proportional to product of flux and speed. Flux remains constant in a shunt machine and it is assumed that during the notching operation. Thus the back emf $E_1 = E_2$.

i.e.　　$I_1 R_2 = I_2 R_1$

$$\frac{R_2}{R_1} = \left(\frac{I_2}{I_1}\right) = \alpha \qquad \text{... (2.11)}$$

Similarly,

$$\frac{R_3}{R_2} = \frac{R_4}{R_2} = \dots = \frac{R_n}{R_{n-1}} = \frac{R_{n+1}}{R_n}$$

$\therefore$　We can write,

$$\frac{I_2}{I_1} = \frac{R_2}{R_1} = \frac{R_3}{R_2} = \frac{R_4}{R_3} = \dots = \frac{R_n}{R_{n-1}}$$

$$= \frac{R_{n+1}}{R_n} = \alpha \qquad \text{... (2.12)}$$

From eq. 2.12,

$$\alpha^n = \frac{R_2}{R_1} \times \frac{R_3}{R_2} \times \frac{R_4}{R_3} \times \dots \times \frac{R_n}{R_{n-1}} \times \frac{R_{n+1}}{R_n}$$

$$= \frac{R_{n-1}}{R_1} = \frac{r_m}{R_1} \qquad \text{... (2.13)}$$

$$R_{n+1} = r_m$$

as

Hence　　$$\alpha = \left(\frac{r_m}{R_1}\right)^{1/n} \qquad \text{... (2.14)}$$

$$R_2 = \alpha R_1$$
$$R_3 = \alpha R_2 = \alpha^2 R_1$$
$$R_4 = \alpha R_3 = \alpha^3 R_1$$

$$\dots\dots\dots\dots\dots\dots\dots$$

$$R_n = \alpha R_{n-1} = \alpha^{n-1} R_1$$
$$r_m = \alpha R_n = \alpha^n R_1$$

∴ The resistance of different sections are :

$$r_1 = R_1 - R_2 = (1 - \alpha)R_1$$
$$r_2 = R_2 - R_3 = \alpha(1 - \alpha)R_1 = \alpha r_1$$
$$r_3 = R_3 - R_4 = \alpha^2(1 - \alpha)R_2 = \alpha^2 r_1$$
$$r_n = R_n - R_{n+1} = \alpha^{n-1}(1 - \alpha)R_1 = \alpha^{n-1}r_1$$

When the upper current limit is fixed, Eq. may be written as

$$\alpha = \left(\frac{r_m}{R_1}\right)^{\frac{1}{n}} = \left(\frac{r_m}{V/I_1}\right)^{\frac{1}{n}} = \left(\frac{I_1 r_m}{V}\right)^{\frac{1}{n}} \quad \ldots (2.15)$$

When the lower current limit is fixed eq. is written as :

$$\alpha = \left(\frac{I_1 r_m}{V}\right)^{\frac{1}{n}} = \left(\frac{r_m}{V} \cdot \frac{I_2}{\alpha}\right)^{\frac{1}{2}} = \left(\frac{r_m I_2}{V}\right)^{\frac{1}{n+1}} \quad \ldots (2.16)$$

Example 2.2 : *A 250 V, 370 kW, d.c. shunt motor has to exert a maximum torque of 150 per cent of full load torque during the starting period. The armature resistance is 0.2 Ω and the full load efficiency is 84 per cent. The number of studs is 8. Find.*

(i) The upper and lower limits of current during starting

(ii) the resistance of each section.

Solution : (i) Full load current = $\dfrac{37 \times 10^3}{250 \times 0.84}$ = 176 A.

Torque is proportional to (ϕI) and the flux is constant, therefore, the torque us proportional to I.

Now maximum torque = 1.5 × full load torque.

∴ Maximum current I_1 = 1.5 × full load current = 264A.

$$\alpha = \left(\frac{r_m I_1}{V}\right)^{\frac{1}{n}} = \left(\frac{0.2 \times 264}{250}\right)^{\frac{1}{7}} = 0.8.$$

Lower limit of current

$$I_2 = \alpha I_1 = 0.8 \times 264 = 211 \text{ A}$$

∴ The current fluctuates between 211 and 264 A during starting.

(ii) Total resistance at starting

$$R_1 = \frac{V}{I_1} = \frac{250}{264} = 0.947 \ \Omega.$$

The resistance of 7 sections are :

$r_1 = (1 - \alpha)\,R_1$	= $(1 - 0.8) \times 0.947$	= 0.189 Ω.
$r_2 = \alpha r_1$	= 0.8×0.189	= 0.151 Ω.
$r_3 = \alpha r_2$	= 0.8×0.151	= 0.121 Ω.
$r_4 = \alpha r_3$	= 0.8×0.121	= 0.077 Ω.
$r_5 = \alpha r_4$	= 0.8×0.097	= 0.097 Ω.
$r_6 = \alpha r_5$	= 0.8×0.077	= 0.062 Ω.
$r_7 = \alpha r_6$	= 0.8×0.062	= 0.050 Ω.

Total resistance of elements = 0.747 Ω.

Motor resistance r_m = 0.200 Ω.

Total resistance at starting R_1 = 0.947 Ω.

Example 2.3 : *Find resistant of a 5 section starter for a 7.5 kW, 250 V, 750 r.p.m. d. c. shunt motor form the given below :*

Maximum torque during stating period = 1.5 times full load torque ; full load efficiency = 85.5 per cent. Armature circuit copper loss is 50 per cent of total loss. Field current = 2.6 A. Find the speed at each stud when the notching takes place.

Solution : Full load line current = $\dfrac{7.5 \times 10^3}{0.855 \times 250}$ = 35.1 A.

Full load armature current = 35.1 − 2.6 = 32.5 A.

Maximum armature current I_1 = 1.5 × 32.5 = 48.8 A.

Hence, total input = $7.5 \times \dfrac{10^{-3}}{0.855}$ = 8770 W

Total losses at full load = 8760 − 7500 = 1270.

Copper loss at full load = $\dfrac{1}{2} \times 1270$ = 635 W

∴ Armature circuit resistance $r_m = \dfrac{635}{(32.5)^2}$ = 0.601 Ω.

Now number of sections n = 5.

$$\therefore \quad \alpha = \left(\frac{r_m I_1}{V}\right)^{\frac{1}{n}} = \left(\frac{0.602 \times 48.8}{250}\right)^{\frac{1}{5}} = 0.652$$

Total resistances at different studs are :

R_1	$= \dfrac{V}{I_1} = \dfrac{250}{48.8}$	= 5.12 Ω.
R_2	$= \alpha R_1 = 0.652 \times 5.12$	= 3.33 Ω.
R_3	$= \alpha^2 R_1 = 0.652^2 \times 5.12$	= 2.18 Ω.
R_4	$= \alpha^3 R_1 = 0.652^3 \times 5.12$	= 1.42 Ω.
R_5	$= \alpha^4 R_1 = 0.652^4 \times 5.12$	= 0.94 Ω.
R_6	$= r_m$	= 0.601 Ω.

The resistances of different sections are :

$$r_1 = R_1 - R_2 = 5.12 - 3.33 = 1.79 \ \Omega$$
$$r_2 = R_2 - R_3 = 3.33 - 2.18 = 1.15 \ \Omega$$
$$r_3 = R_3 - R_4 = 2.18 - 1.42 = 0.76 \ \Omega.$$
$$r_4 = R_4 - R_5 = 1.42 - 0.94 = 0.48 \ \Omega.$$
$$r_5 = R_5 - R_6 = 0.94 - 0.601 = 3.34 \ \Omega.$$

Minimum value of current

$$I_2 = \alpha I_1 = 0.652 \times 48.8 = 31.8 \text{ A}$$

When notching takes place, the current is at its minimum value

∴ Back emf at stud 1 when notching takes place

$$E_1 = V - I_2 R_1 = 250 - 31.8 \times 5.12 = 87.2 \text{ V}.$$

Similarly
$$E_2 = 250 - 31.8 \times 3.33 = 144.1 \text{ V}.$$
$$E_3 = 250 - 31.8 \times 2.18 = 180.6 \text{ V}.$$
$$E_4 = 250 - 31.8 \times 1.42 = 204.8 \text{ V}.$$
$$E_5 = 250 - 31.8 \times 0.94 = 220.1 \text{ V}.$$

When the motor is running at full load the speed is 750 r.p.m. and the back emf.

$$E = 250 - 32.8 \times 0.601 = 230.3 \text{ V.}$$

The back emf is directly proportional to speed. Therefore, the speed at various studs is :

$$\text{Stud 1,} \quad \text{Speed} = 750 \times \frac{E_1}{E} = 750 \times \frac{87.2}{230.2} = 284 \text{ r.p.m.}$$

$$\text{Stud 2,} \quad \text{Speed} = 750 \times \frac{E_2}{E} = 750 \times \frac{144.1}{230.2} = 469 \text{ r.p.m.}$$

$$\text{Stud 3,} \quad \text{Speed} = 750 \times \frac{180.6}{230.2} = 588 \text{ r.p.m.}$$

$$\text{Stud 4,} \quad \text{Speed} = 750 \times \frac{204.8}{230.2} = 667 \text{ r.p.m.}$$

$$\text{Stud 5,} \quad \text{Speed} = 750 \times \frac{220.1}{230.2} = 717 \text{ r.p.m.}$$

2. Starters for DC Series Motor

Referring to Fig. 2.1, for a series motor, we find that on switching at stud 1, there is no back emf and therefore the current is limited by resistance R_1. This current is the maximum current and is given by

$$I_1 = \frac{V}{R_1}$$

The motor now picks up speed and the back emf is developed. Due to the back emf the current falls and its value comes down to I_2 before notching. Just before notching at stud 1, let us say that the back emf, is E_1'.

$$\therefore \quad I_2 = \frac{V - E_1'}{R_1}$$

The current is I_2 and therefore the flux is ϕ_2. The back emf E_1' is proportional to product of speed and flux ϕ_2. Let the speed at that instant be N_1.

$$\therefore \quad E_1' = K\phi_2 N_1 \text{ where K is a constant.}$$

$$I_2 = \frac{V - E\phi_2 N_1}{R_1} \qquad \ldots (2.17)$$

The resistance r_1 of the first section is cut out and the current rises to I_1 and the flux changes to ϕ_1. Let the back emf be E_2''.

$$\therefore \quad E_2'' = K\phi_1 N_1.$$

(speed remains same)

The resistance in the circuit is now R_2.

$$\therefore \quad I_1 = \frac{V - E_2''}{R_2} = \frac{V - K\phi_1 N_1}{R_2} \qquad \ldots (2.18)$$

From Eqn. 2.17, we have,

$$KN_1 = \frac{V - I_2 R_1}{\phi_2}$$

Substituting this in Eqn. 2.18,

$$I_1 = \frac{V - \dfrac{\phi_1}{\phi_2}(V - I_2 R)}{R_2}$$

From Eqn. 2.17, we have,

$$KN_1 = \frac{V - I_2 R_1}{\phi_2}$$

Substituting this in Eq. 2.18,

$$I_1 = \frac{V - \dfrac{\phi_1}{\phi_2}(V - I_2 R)}{R_2}$$

or $\quad R_2 = \dfrac{I_2 \phi_1}{I_1 \phi_2} R_2 - \dfrac{V}{I_1}\left(\dfrac{\phi_1}{\phi_2} - 1\right)$

$$= \alpha\beta R_1 - \frac{V}{I_1}(\beta - 1).$$

In general, for any step R_n,

$$R_n = \alpha\beta R_{n-1} - \frac{V}{I_1}(\beta - 1) \qquad \ldots (2.19)$$

Writing $\sigma = \alpha\beta$, we have

$$R_2 = \delta R_1 - \frac{V}{I_1}(\beta - 1)$$

and $\quad R_n = \delta R_{n-1} - \dfrac{V}{I_1}(\beta - 1).$

The relation between r_m and R_1 is :

$$r_m = R_1 - \left[R_1(1 - \delta) + \frac{V}{I_1}(\beta - 1)\right]\left(\frac{1 - \delta^n}{1 - \delta}\right) \ldots (2.20)$$

But $\quad \dfrac{V}{I_1} = R_1$, so that

$$r_m = R_1\left[1 - \left(\frac{1 - \delta^m}{1 - \delta}\right)(\beta - \delta)\right] \qquad \ldots (2.21)$$

The resistances of sections are :

$$r_1 = R_1 - R_2 = R_1\beta(1 - \alpha) = R_1(\beta - \delta)$$

$$r_2 = R_2 - R_3 = \alpha\beta(R_1 - R_2) = \delta r_1$$

$$r_3 = R_2 - R_4 = \alpha\beta(R_2 - R_3) = \delta^2 r_1$$

....

$$r_n = R_n - R_{n+1} = \alpha\beta(R_{n-1} - R_n) = \delta^{n-1} r_1$$

Example 2.4 : *Estimate the number of resistance sections and the resistance of each section for the starter of a 7.5 kW, 460 V, d.c. series motor. The starting current varies from 1.5 to 2 times full load current.*

The efficiency is 80 per cent and the resistance of machine measured between terminals 1.8 Ω. Assume that the flux increases by 10 per cent as the current rises from 1.5 to 2 times rated full load current.

Solution : Full load current

$$= \frac{7500}{460 \times 0.8} = 20.3 \text{ A.}$$

$$I_1 = 2 \times 20.3 = 40.6 \text{ A}$$

and $\quad I_2 = 1.5 \times 20.3 = 30.4 \text{ A}$

$$\alpha = \frac{I_2}{I_1} = \frac{30.4}{40.6} = 0.75$$

and $\quad \beta = \frac{\phi_1}{\phi_2} = 1.1$

$\therefore \quad \delta = \alpha\beta = 0.75 \times 1.1 = 0.825$

and $\quad R_1 = \frac{V}{I_1} = \frac{460}{40.6} = 11.3 \text{ }\Omega.$

The resistances of various sections are :

$r_1 = R_1 (\beta - \delta) \qquad = 11.3(1.1 - 0.825) \qquad = 3.10 \text{ }\Omega.$

$r_2 = \delta r_1 \qquad\qquad = 0.825 \times 3.1 \qquad = 2.56 \text{ }\Omega.$

$r_3 = \delta^2 r_1 \qquad\qquad = 0.825^2 \times 3.1 \qquad = 2.11 \text{ }\Omega.$

$r_4 = \delta^3 r_1 \qquad\qquad = 0.825^3 \times 3.1 \qquad = 1.74 \text{ }\Omega.$

Adding these resistances, we have the total resistance equal to 9.51 Ω. To this if we add the value of motor resistance, i.e. 1.8 Ω, we get a total resistance of 11.31 Ω. This value is equal to R_1.

$\therefore$ We should not increase the resistance steps beyond 4 otherwise the total resistance would become higher than the required.

$\therefore$ Number of sections = 4 and number of studs = 5.

Example 2.5 : *Find the resistance of each of the four sections if a starter for a 500 V series crane motor having a resistance of 0.3 Ω to give a maximum current of 100 A during starting. Assume the magnetization curve to be a straight line passing through origin.*

Solution : Total resistance at starting

$$R_1 = \frac{V}{I_1} = \frac{500}{100} = 5\Omega$$

Now, $\quad \alpha = \frac{I_2}{I_1}$ and $\beta = \frac{\phi_1}{\phi_2}$

But, $\quad \frac{\phi_1}{\phi_2} = \frac{I_1}{I_2}$ as the magnetization curve is linear.

$\therefore \quad \beta = \frac{\phi_1}{\phi_2} = \frac{I_1}{I_2}$

hence, $\quad \delta = \alpha\beta = \left(\frac{I_2}{I_1}\right) \times \left(\frac{I_1}{I_2}\right) = 1$

The resistances of the four sections are :

$r_1 = (\beta - \delta)R_1 = (\beta - 1)R_1$

$r_2 = \delta r_1 = r_1$

$r_3 = \delta^2 r_1 = r_1$

$r_4 = \delta^3 r_1 = r_1$

Now, $r_1 + r_2 + r_3 + r_4 + r_m = R_1 = 5.0 \text{ }\Omega.$

$\therefore r_1 + r_2 + r_3 + r_4 = 5.0 - 0.3 = 4.7 \text{ }\Omega$

But $\qquad r_1 = r_2 = r_3 = r_4$

$$r_1 = r_2 = r_3 = r_4 = \frac{4.7}{4} = 1.175 \text{ }\Omega$$

3. Starters for Three Phase Slipring Induction Motor

Let us neglect the resistance and leakage reactance of the stator winding. Let X be the standstill reactance and E be the standstill emf of the rotor winding. We now refer to Fig. 2.1.

For Stud 1 :

The motor is switched on to the supply. The machine draws the maximum current I_1. Let the slip be s_1.

$\therefore \qquad I_1 = \dfrac{s_1 E}{\sqrt{(R_1^2 + s_1^2 X^2)}}$

The machine picks up speed and the slip falls and so does the current. The value of current just before notching is I_2. Let the slip be s_1 at this instant.

$\therefore \qquad I_2 = \dfrac{s_2 E}{\sqrt{(R_1^2 + s_2^2 X^2)}}$

Stud 2 : The resistance r_1 is cut out and the resistance in the circuit is R_2.

The current rises to its maximum value I_1. Let us consider that there is no change in speed during notching operation.

$\therefore$ The slip just after notching is s_2.

Hence, $\qquad I_1 = \dfrac{s_2 E}{\sqrt{(R_2^2 + s_2^2 X^2)}}$

Just before notching at stud 2, the current falls to I_2 and slip to s_3.

$\therefore \qquad I_2 = \dfrac{s_1 E}{\sqrt{(R_1^2 + s_1^2 X^2)}} = \dfrac{s_2 E}{\sqrt{(R_2^2 + s_2^2 X^2)}} = ...$

$$= \dfrac{s_{n+1} E}{\sqrt{(r_m^2 + s_{n+1}^2 X^2)}} \qquad ... (2.22)$$

From Eq. 2.22

$$\frac{R_1}{s_1} = \frac{R_2}{s_2} = ... = \frac{r_m}{s_{n+1}} \qquad ... (2.23)$$

and $\qquad I_2 = \dfrac{s_2 E}{\sqrt{(R_1^2 + s_2^2 X^2)}} = \dfrac{s_3 E}{\sqrt{(R_2^2 + s_3^2 X^2)}} = ...$

$$= \dfrac{s_{n+1} E}{(R_n^2 + s_{n+1}^2 X^2)} \qquad ... (2.24)$$

From Eq. 2.24

$$\frac{R_1}{S_2} = \frac{R_2}{S_3} = \ldots - \frac{R_n}{S_{n+1}} \qquad \ldots (2.25)$$

Combining eq. (2.23) and (2.25)

$$\frac{S_2}{S_1} = \frac{S_3}{S_2} = \ldots = \frac{S_{n+1}}{S_n} = \frac{R_2}{R_1} = \frac{R_3}{R_2} = \ldots = \frac{r_m}{R_n} = \alpha \quad \ldots (2.26)$$

Now, slip at starting s1 = 1 and slip at $(n + 1)^{th}$ stud, S_{n+1} = slip at full load = s_m. (This is true provided the upper limit of current is the full load current).

$$\therefore \quad \frac{S_2}{S_1} = \frac{S_3}{S_2} = \ldots = \frac{S_m}{S_n} = \frac{R_2}{R_1} = \frac{R_3}{R_2} = \ldots\ldots = \frac{r_m}{R_n} = \alpha$$

or $\qquad s_m = (\alpha)^n$

$$\alpha = (s_m)^{1/n} \qquad \ldots (2.27)$$

Also $\qquad \dfrac{r_m}{R_1} = (\alpha)^n$

or $\qquad r_m = R_1(\alpha)^n \qquad \ldots (2.28)$

Further, $\quad R_2 = \alpha R_1$

$$R_3 = \alpha R_2 = \alpha^2 R_1$$

$$R_4 = \alpha R_3 = \alpha^3 R_1$$

$$\cdots\cdots\cdots\cdots\cdots\cdots$$

$$R_n = \alpha R_{n-1} = \alpha^{n-1} R_1$$

The sections are :

$$r_1 = R_1 - R_2 = R_1(1 - \alpha)$$

$$r_2 = R_2 - R_3 = R_1(\alpha - \alpha 2) = \alpha r_1$$

$$r_3 = R_3 - R_4 = R_1(\alpha^2 - \alpha^3) = \alpha^2 r_1$$

$$\cdots\cdots\cdots\cdots\cdots\cdots\cdots$$

$$r_n = R_n - R_{n+1} = R_1(\alpha^{n-1} - \alpha^n) = \alpha^{n-1} r_1$$

From above eq.

$$\left(\frac{I_2}{I_1}\right)^2 = \frac{s_2^2(R_1^2 + s_1^2 X^2)}{s_1^2(R_1^2 + s_2^2 X^2)} = \frac{s_2^2(R_1^2 + X^2)}{(R_1^2 + s_2^2 X^2)}$$

$$= \frac{R_1^2 + X^2}{\left(\dfrac{R_1}{\alpha}\right)^2 + X^2} = \alpha^2 \text{ for normal motors}$$

or $\qquad \dfrac{I_2}{I_1} = \alpha$ for normal motors $\qquad \ldots (2.29)$

The ratio of the rotor current limits is, therefore, approximately equal to α and this is also the approximate ratio of stator currents.

Example 2.6 : *Design the sections of a rotor starter for a 75 kW, 3 phase induction motor, using 7 notches. Rotor resistance per phase = 0.018 Ω. The upper current limit is to be full load current, for which the slip is 2 per cent.*

Solution : Notches or studs = 7.

$\qquad \therefore \quad$ Numbers of resistance elements = n = 6.

$$\alpha = (s_m)^{1/n} = (0.02)^{1/6} = 0.521.$$

$$r_m = R_1(\alpha)^n = R_1 s_m$$

$$\therefore \qquad R_1 = \frac{r_m}{s_m} = \frac{0.018}{0.2} = 0.9$$

The resistances of elements are :

$r_1 = r_1 (1 - \alpha)$	$= 0.9(1 - 0.521)$	$= 0.432\ \Omega$
$r_2 = \alpha r_1$	$= 0.521 \times 0.432$	$= 0.225\ \Omega$
$r_3 = \alpha^2 r_1$	$= 0.521^2 \times 0.432$	$= 0.108\ \Omega$
$r_4 = \alpha^3 r_1$	$= 0.521^3 \times 0.432$	$= 0.062\ \Omega$
$r_5 = \alpha^4 r_1$	$= 0.521^4 \times 0.432$	$= 0.032\ \Omega$
$r_6 = \alpha^5 r_1$	$= 0.521^5 \times 0.432$	$= 0.017\ \Omega$
r_m		$= 0.018\ \Omega$
Total resistance of rotor		$= 0.894\ \Omega$

This resistance is nearly equal to R_1.

Example 2.7 : *Design the 6 sections of 7 stud rotor starter for a 3 phase wound rotor induction motor. The slip at full load current is 2% and the maximum starting current is 1.5 times full load current. The resistance of rotor per phase is 0.02 Ω.*

Solution : Rotor current $= \dfrac{E}{\sqrt{\left(\dfrac{R}{s}\right)^2 + X^2}}$

$$= \frac{sE}{R} \text{ if rotor reactance is neglected.}$$

This means that slip changes directly with rotor current. The full load slip is 2%.

$\qquad \therefore \quad$ Slip at 1.5 times full load current = 1.5 × 2 = 3%

$\qquad$ From Eq. 2.27,

$$\alpha = (s_m)^{1/n} = (0.03)^{1/6} = 0.557$$

$$R_1 = \frac{r_m}{s_m} = \frac{0.02}{0.03} = 0.666\ \Omega$$

The resistances of the sections are :

$r_1 = (1 - \alpha)R_1$	$= 1 - (0.557) \times 0.666$	$= 0.295\ \Omega$
$r_2 = \alpha r_1$	$= 0.557 \times 0.295$	$= 0.165\ \Omega$
$r_3 = \alpha^2 r_1$	$= 0.557^2 \times 0.295$	$= 0.092\ \Omega$
$r_4 = \alpha^3 r_1$	$= 0.557^3 \times 0.295$	$= 0.051\ \Omega$
$r_5 = \alpha^4 r_1$	$= 0.557^4 \times 0.295$	$= 0.028\ \Omega$
$r_6 = \alpha^5 r_1$	$= 0.557^5 \times 0.295$	$= 0.016\ \Omega$
r_m		$= 0.020\ \Omega$
Total resistance of rotor		$= 0.667\ \Omega$

This value is almost equal to R_1.

Example 2.8 : *A 37.5 kW, 400 V, 50 Hz, 3 phase slip ring induction motor has a copper loss of 1200 W per phase at full load and a friction and windage loss of 500 W. The rotor resistance per phase is 0.15 Ω*
Assuming that starting current is not to exceed 1.25 times full load current, work out the steps of a 4 section starter.

Solution : Gross rotor output = 37.5 kW = 37500 W

Rotor input = 37500 + 1200 + 500 = 39200 W

Now   slip $= \dfrac{\text{rotor copper loss}}{\text{rotor input}}$

slip at full load

$$s_m = \frac{1200}{39200} = 0.0306$$

Slip at 1.25 times full load current = 1.25×0.0306
$$= 0.03825$$

Now, $\alpha = (s_m)^{1/n} = (0.03825)^{1/4}$

and $R_1 = \left(\dfrac{r_m}{s_m}\right) = \dfrac{0.15}{0.3825} = 3.92\ \Omega$

The resistances of sections are :

$r_1 = R_1(1-\alpha)$	$= 3.92(1-0.442)$	$= 2.190\ \Omega$
$r_2 = \alpha r_1$	$= 0.442 \times 2.19$	$= 0.968\ \Omega$
$r_3 = \alpha^2 r_1$	$= (0.444)^2 \times 2.19$	$= 0.428\ \Omega$
$r_4 = \alpha^3 r_1$	$= (0.442)^3 \times 2.19$	$= 0.189\ \Omega$
r_m		$= 0.150\ \Omega$
Total resistance of rotor		$= 3.925\ \Omega$

This is almost equal to R_1.

2.4 DESIGN OF FIELD REGULATORS FOR DC MACHINES

2.4.1 Shunt Generators

Calculation of steps of field regulator of a shunt generator is explained in the following example.

Example 2.9 : *Find the section resistance of a 7 stud filed regulator for a generator to give the limits of 500 and 560 V in equal steps. The magnetization curve is given in Fig. 2.2. The field is 934 Ω*

Solution :

There are 7 studs and therefore the number of resistance sections is 6. The range 500 to 560 V is divided into 6 sections of 10 V each. The method for calculation of resistances is shown as under :

Take the step at 550 V. The field current corresponding to 550 V is 0.55 A.

$\therefore$ Field circuit resistance $= \dfrac{550}{0.55} = 1000$

$\therefore$ Resistance to be inserted = 1000 – 934 = 66 Ω

Resistance of section = 66 Ω

Take the step at 540 V. The field current is 0.51 A.

$\therefore$ Total field circuit resistance $= \dfrac{540}{0.51} = 1058\ \Omega$

Total resistance to be inserted = 1058 – 934 = 124 Ω

Resistance of section = 124 – 66 = 58 Ω

The value of resistances of other sections can be similarly calculated. The results are tabulated below :

Stud No.	Emf V	Field Current A	Total Resistance W	Section Resistance W
1	560	0.60	934	–
2	550	0.55	1000	66
3	540	0.51	1058	58
4	530	0.48	1102	44
5	520	0.455	1142	40
6	510	9.435	1172	30
7	500	0.4175	1195	23

2.4.2 Shunt Motor

The method for calculation of resistance elements of a field regulator for shunt motor is illustrated in the following example :

Example 2.10 : *Find the suitable shunt field regulator resistance elements for a speed range of 750 to 1350 r.p.m. in increments of 150 r.p.m. for a 250 V shunt motor. Field resistance = 100 Ω. Fig. 2.3 shows the O.C.C. for 750 r.p.m.*

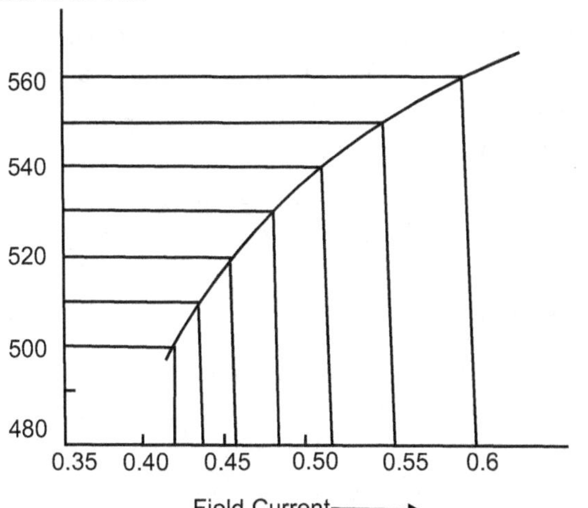

Fig. 2.2 : Magnetizing curve of Example 2.9

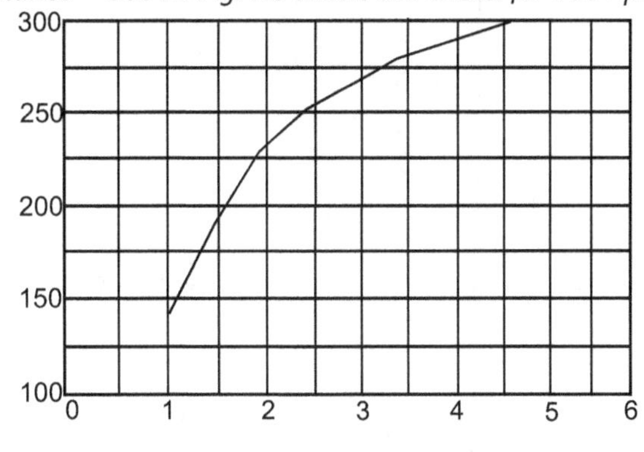

Fig. 2.3 : Magnetization curve of Example 2.10

Solution : Speed range = 750 tp 1350 r.p.m.

Number of selections = $\dfrac{1350 - 750}{150}$ = 4.

Stud 1 : Speed = 750 r.p.m.

Voltage = 250 V.

Corresponding to 250 volts, field current = 2.5 A

Shunt field circuit resistance = $\dfrac{250}{2.5}$ = 100 Ω.

Shunt field resistance = 100 Ω.

∴ External resistance = 0.

Stud 2 : Speed 900 r.p.m.

In order to get this speed the flux must be reduced to $\dfrac{750}{900}$ = 0.833 of the flux with 750 r.p.m. or the voltage on O.C.C. should be 250 × 0.833 = 208 V.

Corresponding to 208 V, field current = 1.8 A.

∴ Shunt field circuit resistance = $\dfrac{250}{1.8}$ = 139 Ω

∴ External resistance = 139 − 100 = 39 Ω.

Resistance of step = 39 Ω.

The rest of the calculations are tabulated below :

N Speed r.p.m.	$E_{mf} = \dfrac{250 \times 750}{N}$ V	Field Current I_f from O.C.C. A	Field Circuit Resistance $R_l = 250 / I_f$ Ω	Rheostat Resistance $R = R_l - 100$ Ω	Section Resistance Ω
750	250	2.5	100	0	...
900	208	1.8	139	39	39
1050	178	1.4	179	79	40
1200	156	1.2	208	108	29
1350	139	1.05	238	138	30

2.4.3 Potentiometer Regulators

This regulators are used when a wide variation in voltage or speed control is desired. This is very beneficial if a gradual control is desired or when the field current is small.

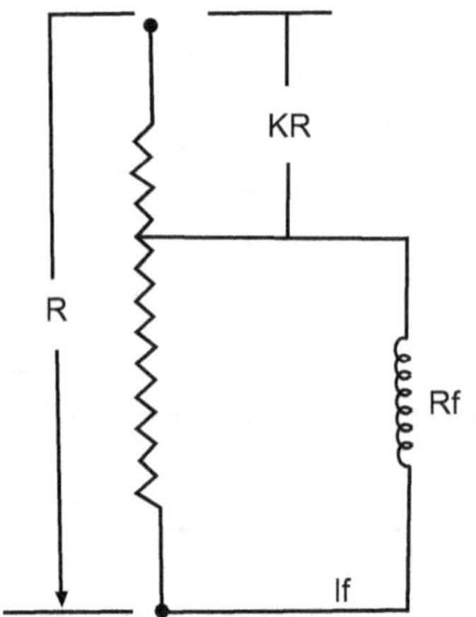

Fig. 2.4 : Potentiometer regulator

Let V = voltage applied to potentiometer,

R = total resistance of potentiometer

KR = resistance of untapped portion of potentiometer,

R_f = field resistance, m = $\dfrac{R}{R_f}$

I = total current and I_f = field current.

Total resistance between terminal is

$$= KR + \frac{R_f \times (1 - K)R}{R_f + (1 - K)R} = \frac{(K - K^2)R^2 + R_fR}{R_f + (1 - K)R}$$

$$= \frac{(K - K^2)m^2R_f^2 + mR_f^2}{R_f + mR_f(1 - K)}$$

$$= R_f\left[\frac{(K - K^2)m^2 + m}{1 + m(1 - k)}\right]$$

Total current following

$$I = \frac{V}{R_f}\left[\frac{(1 - K)m + 1}{(K - K^2)m^2 + 1}\right] \qquad \ldots (2.30)$$

Field current

$$I_f = \frac{V}{R_f}\left[\frac{(1 - K)}{(K - K^2)m + 1}\right] \qquad \ldots (2.31)$$

The maximum field current is $\dfrac{V}{R_f}$.

The value of total current I and the field current I_f can be found for any setting of the potentiometer (i.e. for any value of K) if the ratio m = R/R_f is specified.

2.5 DESIGN OF RESISTANCE ELEMENTS OF FIELD REGULATORS

The design of field regulators for both motors and generators follows the same procedure. The method of calculation of resistance of the field regulators have been designed in the following article.

2.5.1 Material for Resistance Elements

The material usually used is constant an which has a resistivity, 0.46 to 0.53 Ω/m and mm^2 a resistance temperature coefficient, 0.00002/°C (virtually zero); and a maximum operating temperature, 500°C.

Nichrome can also be used. It has a resistivity of 1.1 to 1.27 Ω/m and mm^2 and a maximum operating temperature of 900 − 1000°C.

2.5.2 Size of Wire

The field rheostats of generators and motors are designed to for continuous duty. They are mostly wire wound units ranging in capacity from a few watt a few kilowatt.

The size of wire for a particular current carrying capacity can be calculated as under :

Let, I = current carrying capacity of wire, A;

d = diameter of wire, mm;

l = length of wire, m ;

ρ = resistivity of wire material, Ω/m and mm^2;

λ = specific heat dissipation, $W/m^2 - °C$;

and θ = maximum permissible temperature rise, °C.

$\therefore$ Loss to be dissipated = $I^2 R$

$$= I^2 \rho = \frac{l}{(\pi/4)\, d^2} \qquad ...(2.32)$$

Dissipating surface of wire

$$S = \pi dl \times 10^{-3}$$

$$\text{Loss dissipated} = S\lambda\theta = (\pi dl \times 10^{-3})\, \lambda\theta \qquad ...(2.33)$$

Equating (2.32) and (2.33), we have,

$$I = \left(\frac{\pi^2}{4}\frac{\lambda\theta}{\rho} \times 10^{-3}\right)^{1/2} d^{1.5} \qquad ...(2.34)$$

$$= K\, d^{1.5} \qquad ...(2.35)$$

where $K = \left(\dfrac{\pi^2}{4}\dfrac{\lambda\theta}{\rho} \times 10^{-3}\right)^{1/2} \qquad ...(2.36)$

The temperature rise θ may be taken as between 100 to 150°C and specific heat dissipation, λ, as 20 $W/m^2 - °C$.

For constantan

ρ = 0.5 Ω/m and mm^2. Taking θ = 100°C, we have :

$$K = \sqrt{\frac{\pi^2}{4} \times \frac{20 \times 100}{0.5} \times 10^{-3}} = 3.2.$$

Hence $I = 3.2\, d^{1.5} \qquad ...(2.37)$

For $\theta = 150°C$, we have $I = 3.9\, d^{1.5} \qquad ...(2.38)$

For nichrome

ρ = 1.2 Ω/m and mm^2. For θ = 100°C, we have

$$K = \sqrt{\frac{\pi^2}{4} \times \frac{20 \times 100}{1.2} \times 10^{-3}} \approx 2$$

Hence, $I = 2\, d^{1.5} \qquad ...(2.39)$

For $\theta = 150°C$, we have $I = 2.5\, d^{1.5} \qquad ...(2.40)$

The above relationship are based upon a specific heat dissipation of 20 $W/m^2 - °C$.

From Eq. (2.37) to (2.40), the diameter of wire can be found out. The standard wire size is selected by referring to standard specification (British standard sizes of wires).

The resistance per unit length can be worked out and the length of wire to be used for each resistance section can be calculated.

2.5.3 Resistance Box

Regulator and starter rheostats having heavy wire (20 SWG and under or diameter about 1 mm and over) can be arranged in the form of a spiral supported at the ends by porcelain cleats or studs on an iron plate. Smaller gauges must be wound on supporting ceramic tubes. The rheostat framework is enclosed in a ventilated cast iron box. Fig. 2.5 shows the design of a box suitable for arranging the wires in open spirals.

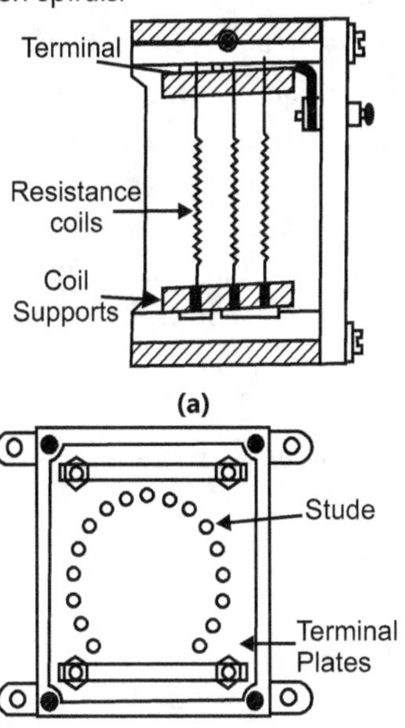

Fig. 2.5 : Resistance box of field regulator

Example 2.11 : *Design the resistance elements of field regulator of the generator of example 2.9. The resistance of the 6 sections are 66, 58, 44, 40, 30, 23 Ω. The maximum current is 0.6 A.*

Solution : The field regulator is designed for the maximum current i.e. 0.6 A.

(i) Using constantan wire, a temperature rise of 150°C and a dissipation of 20 $W/m^2 - °C$, we have :

$$I = 3.9\, d^{1.5} \text{ or } 0.6 = 3.9\, d^{1.5}$$

$$\therefore \qquad d = 0.288 \text{ mm.}$$

From standard specification, the nearest standard conductor available is 31 SWG and it has a diameter of 0.2946 mm and an area of 0.06819 mm^2.

Resistance per metre length = ρ/α = 0.5/0.06819 = 7.35Ω

Length of wire in section No. 1 = 66/7.35 = 8.96 m.

Length of wire in section No. 2 = 58/7.35 = 7.88 m.

Length of wire in section No. 3 = 44/7.35 = 5.97 m.

Length of wire in section No. 4 = 40/7.35 = 5.34 m.

Length of wire in section No. 5 = 30/7.35 = 4.07 m.

Length of wire in section No. 6 = 23/7.35 = 3.12 m.

2.6 DESIGN OF CHOKES

2.6.1 Design Procedure

The design of chokes is very much similar to design of small transformers.

Let V = voltage applied across the choke, V;

I = current through the choke, A;

Q = volt ampere rating of choke =– VI;

T_e = turns per volt ;

A_i = area of core, m^2 ;

and A_ω = window area, m^2.

It should be noted that the volume of copper in a choke is half of that of corresponding transformer and therefore a higher value of T_e. Actually the choke can be thought of as a transformer with 1 : 1 turns ratio and with an output of $Q/2$. Thus the value of T_e corresponding to a volt-ampere rating of $Q/2$.

Flux $\phi_m = 1.44\, f\, T_e$

The value of maximum flux density is taken as about 1 Wb/m^2 for the core.

Net area of core

$$A_i = \phi_m/B_m$$

Assuming a stacking factor of 0.9, gross core area,

$$A_{gi} = A/0.9 = 1.11\, A_i$$

E – 1, T – U or J type laminations are usually used.

A square section is normally used for the central limb as a square former is easy to manufacture and a square section gives a smaller length of mean turn as compared to the one given by rectangular section.

$\therefore$ Approximate width of central limb $A = \sqrt{A_{gi}}$

Winding

Number of turns

$$T = V \times T_e$$

Area of wire $a = I/\delta$ where δ = current density.

The value of current density is about 2.5 A/mm^2.

Diameter of bare wire $d = \sqrt{(4/\pi)a}$.

Referring to standard specifications, the nearest standard size of wire is selected. Usually enamelled wire is used but for larger sizes cotton covered wire is used.

Let, d_1 = diameter of insulated wire.

$\therefore$ Space required in the window for the winding = Ta/S_f.

where S_f = space factor = $0.8\,(d/d_1)^2$.

In order to accommodate former etc. this area is increased by about 20 percent.

$\therefore$ Total window area $A_\omega = 1.2\, Ta/S_f$.

Impedance of Choke

$$Z = V/I = T/(T_e I) \qquad \ldots (2.41)$$

Now $I = \dfrac{\text{mmf required for iron parts} + \text{mmf required for air gaps}}{T}$

$\therefore$ $I = \dfrac{at_i l_i + n_g\, at_g\, l_g}{T} \qquad \ldots (2.42)$

where, at_i = mmf (r.m.s.) per metre required for iron parts corresponding to flux density B_m.

(The mmf per metre corresponding to B_m can be read off from the relevant B – 'at' curves. This mmf should be divided by $\sqrt{2}$ to get the value at_i)

$$at_g = 800{,}000\, B_m/\sqrt{2} \qquad \ldots (2.43)$$

l_g = length of each air gap

n_g = number of air gaps in series in the flux path

Hence $Z = \dfrac{T^2}{(at_i\, l_i + n_g\, at_g\, l_g)T_e} \qquad \ldots (2.44)$

Usually the value of mmf required for iron parts is very small as compared with mmf required for air gaps and, therefore we can neglected $at_i l_i$, for practical purposes.

The value of length of l_g can be adjusted to get the required value of impedance.

When E – 1, T – U or J type laminations are used, there are two air gaps in the magnetic path.

Table : 2.1 Turns per volt T_e

VA	Turns per Volt	VA	Turns per volt
10	23.3	200	3.5
15	17.5	250	2.8
20	14.0	300	2.8
25	11.7	400	2.3
50	7.0	500	2.0
75	5.6	750	1.7
100	4.6	750	1.7
150	4.0		

Example 2.12 : *Design an iron cored choke to be connected across a 230 V, 50 Hz supply and to array a current of 4 A.*

Solution :

Core

Apparent power

$$Q = VI = 230 \times 4 = 920 \text{ VA.}$$

The choke may be thought of a 1 : 1 transformer having a rating of $Q/2$ volt ampere.

Corresponding to $920/2 = 460$ VA, turns per volt $T_e = 2$ (from Table 2.1).

Flux in the core

$$\phi_m = \frac{1}{4.44 \times 50 \times 2} = 2.21 \times 10^{-3} \text{ Wb.}$$

The maximum flux density in the core is assumed as 1.0 Wb/m^2.

$\therefore$ Net area of core

$$A_i = 2.21 \times 10^{-3} / 1.0 \text{ m}^2$$
$$= 2210 \text{ mm}^2$$

Gross area of core

$$A_{gi} = 2210/0.9 = 2500 \text{ mm}^2$$

Taking a square section for the central limb, width of central limb, $A = 50$ mm.

Winding :

Number of turns

$$T = V \times T_e = 230 \times 2 = 460$$

Taking a current density of 2.3 A/mm^2 for the winding.

Area of conductor required $a = 4/2.3 = 1.74$ mm^2.

Required diameter of bare conductor = 1.49 mm.

Referring to standard specification (Synthetic enamelled copper conductors) the nearest standard conductor has :

bare diameter

$$d = 1.5 \text{ mm,}$$

diameter with covering

$$d_1 = 1.605 \text{ mm (medium covering)}$$

Area of conductor provided

$$a = 1.76 \text{ mm}^2$$

Space factor

$$S_f = 0.8 \left(\frac{1.5}{1.605}\right)^2 = 0.68.$$

Window area required

$$A_w = 1.2 \frac{T_a}{S_f} = \frac{1.2 \times 460 \times 1.76}{0.68} = 1420 \text{ mm}^2$$

Impedance of Choke

From Eq. 2.43, for

$$B_m = 1 \text{ Wb/m}^2, \text{ we have,}$$

$$at_k = \frac{800000}{\sqrt{2}} B_m = \frac{800000}{\sqrt{2}} \times 1$$

$$= 563000 \text{ A/m.}$$

Number of air gaps in series in the flux path $n_g = 2$.

$\therefore$ From Eq. 2.44, impedance of choke

$$Z = \frac{T^2}{Te(at_i \cdot l_i + n_g \, at_g \, l_g)}$$

Now $Z = 230/4 = 57.5 \ \Omega$, $T = 460$ and $Te = 2$.

Neglecting mmf required for iron parts, we have,

$$57.5 = \frac{(460)^2}{2(2 \times 563000 \, l_g)}$$

Hence length of air gap $l_g = 1.61$ mm.

The impedance of coil is approximately equal to its reactance.

$\therefore$ Inductance of choke coil

$$L = \frac{X}{2\pi f} = \frac{57.5}{2\pi \times 50} \text{ H} = 183 \text{ mH.}$$

2.7 DESIGN OF ELECTROMAGNETS

2.7.1 Introduction

Electromagnets are used for the purposes of pulling, lifting and holding. The large size electromagnets are used for lifting heavy loads while small electromagnets are used for holding of armatures of relays and values. The general principles of operation and design are the same for all types of electromagnets irrespective of their application. However, problems of heat dissipation and insulation, and also constructional details become more complicated as the size of the electromagnet increases.

2.7.2 Types of Electromagnets

Electromagnets are classified into two general categories :

    1.   Tractive type,   2.   Portative type

1. **Tractive Electromagnets :**

   These electromagnets are usually known as solenoids and are designed to produce mechanical part required to do work. Tractive magnets are operated either from a.c. or d.c. supply. Tractive magnets are used in electric bells and buzzers, telegraph apparatus, electric clocks, time stamps, electric pendulums, current limiting magnetic switches and relays, telephone relays magnetic brakes, circuit breakers, track switches and electric valves.

2. **Portative Electromagnets :**

   These electromagnets generally function as holding magnets. They are usually operated from d.c. supply. Portative magnets are used as lifting magnets, magnetic chucks, magnetic clutches, magnetic couplings, magnetic separators ad low voltage releases.

   There are many types of electromagnets but for the sake of brevity, only three commonly used types of electromagnets are described here. These are :

   (i)   Flat-faced armature type,

   (ii)   Horse shoe type, and

   (iii)   Flat-faced plunger type.

2.7.3 Construction of Electromagnets

Electromagnets consist of a ferromagnetic core which carries the flux and a winding which produces a flux when excited by an external source.

1. **Electromagnet Core Materials :**

   Soft magnetic materials are used for construction of core of the electromagnets. Most these materials contain the ferromagnetic elements like iron, nickel and cobalt in various combinations. Sometimes some non-ferro-magnetic elements like silicon, molybdenum and chromium are added to obtain desirable from materials.

2. **Electromagnet Coils :**

   Coils are used in electromagnets as an exiting surface for production of magnetic field. A coil, usually, consists of wire wound like a helical thread to form a layer, there being one or more layers to the coil. Insulation, such as paper is sometimes placed between layers. The usual material for the conductors is usually round except in coils made of heavy wire where a square, or a rectangular section with rounded corners is used.

2.7.4 Design of Magnet Coil

The coil insulation is of the sheet form. It is usually made of cloth or paper treated with varnish, glass, synthetic resin bonded paper and synthetic resin impregnated compressed laminates of wood.

The coils for electromagnets are usually circular in shape and rectangular in cross-section. The various terms connected with circular coils (Fig. 2.6) are explain below :

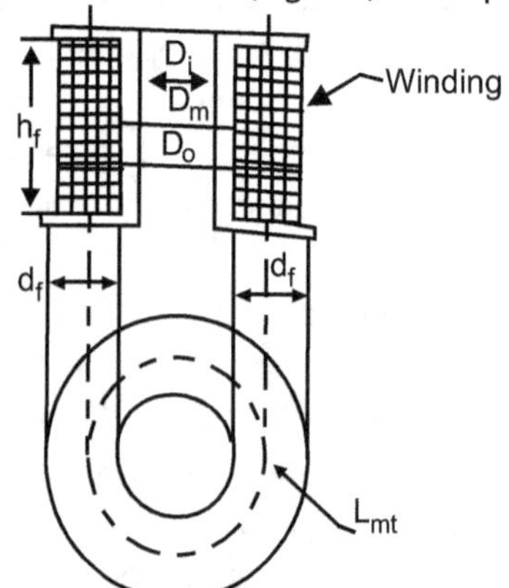

Fig. 2.6 : Electromagnet coil

D_i = inside diameter of coil,
D_a = outside diameter of coil,
D_m = mean diameter of coil,
L_f = radial depth of coil,
h_f = axial length of coil,
L_{mt} = length of mean turn

Mean diameter

$$D_m = \frac{D_i + D_0}{2} = D_i + d_f$$

Outside diameter

$$D_0 = D_i + 2d_f = D_m + d_f$$

Depth of winding

$$d_f = \frac{D_0 - D_i}{2} = D_0 - D_m = D_m - D_i$$

Cross winding area

$$A_\omega = h_f \times d_f$$

Length of mean turn,

$$L_{mt} = \pi D_m = \pi \left(\frac{D_i + D_0}{2}\right)$$
$$= \pi(D_0 - d_f) = \pi(D_i + d_f)$$

Total heat dissipating surface = $2\,L_{mt}\,(d_f + h_f)$

Outer cylindrical heat dissipating surface = $\pi\,D_0 h_f$

Outer and inner cylindrical heat dissipating surface
$$= 2\,L_{mt}\,h_f$$

Coming to the design of electromagnet coils,

Let At = mmf per coil, A;

     V = terminal voltage, V;

     ρ = resistivity of conductor, $\Omega/m/mm^2$;

     δ = current density, A/mm^2;

     L_{mt} = length of mean turn, m;

     I = current, A;

     T = number of turns; R = resistance, Ω;

     a = area of conductor, mm^2;

     S_f = space factor;

     A_ω = total winding section, m^2;

     S = cooling surface of coil, m^2;

     c = cooling co-efficient, $C°\text{-}m^2/W$,

     θ_m = temperature rise of coil surface over ambient medium, °C

The ambient temperature is usually assumed as 20°C.

We have,

$$\text{coil resistance } R = \frac{\text{voltage across the coil}}{\text{current in the coil}} = \frac{V}{I}$$

or area of conductor $a = TI\dfrac{\rho L_{mt}}{V} = AT\dfrac{\rho\,L_{mt}}{V}$

The resistivity of copper at working temperature is :

$$= 0.01734\,[1 + 0.00393$$
$$\text{(temperature rise above 20°C)]}$$

Current : The actual value of current is not known until the conditions of cooling are known. A preliminary value of current is obtained by assuming a suitable current density δ.

$$\therefore \quad I = \delta a \qquad \qquad ...(2.45)$$

Turns : The number of turns are obtained by dividing the total mmf by value of current obtained above.

$$T = \frac{AT}{I} \qquad \qquad ...(2.46)$$

Winding Section : The total area occupied by conductors is equal to the product of number of turns and the area of each conductor.

$$\text{Total winding area } A_w = \frac{\text{number of turns} \times \text{area of each conductor}}{\text{space factor}}$$

$$= \frac{T\,a}{S_f} \text{ mm}^2 = \frac{T\,a}{S_f} \times 10^{-6} \text{ m}^2 \qquad ...(2.47)$$

But $\quad A_w$ = height of winding × depth of winding

$$= h_f \times d_f \text{ m}^2 \qquad \qquad ...(2.48)$$

if both h_f and d_f are expressed in m.

From Eq. 2.47 and 2.48.

$$\left.\begin{aligned} a &= \frac{S_f\, A_w}{T} \times 10^6 \text{ mm}^2 \\[2mm] &= \frac{S_f h_f d_f}{T} \times 10^6 \text{ mm}^2 \end{aligned}\right\} \qquad ...(2.49)$$

Example 2.13 : *A plunger type magnet ha to lift a mass of 200 kg from a distance of 5 mm. The area of pole face is 5×10^{-3} m^2. Find the current required if the exciting coil has 3000 turns. Assume that the mmf required for iron parts = 10 percent of air gap mmf. Neglect fringing.*

Solution : The plunger type magnet exerts force through one pole face only. Therefore, force exerted

$$F = 0.051 \frac{B^2 A}{\mu_0} \text{ kg}$$

or $\quad 200 = \dfrac{0.051\, B^2 \times 5 \times 10^{-2}}{4\pi \times 10^{-7}}$

or flux density in air gap

$$B = 0.003 \text{ Wb/m}^2$$

Mmf required for air = 800,000 $B l_g$

$$= 800,000 \times 0.993 \times 5 \times 10^{-3} = 3972 \text{ A.}$$

Mmf required for iron parts

$$= 0.1 \times 3972 = 397 \text{ A.}$$

Total mmf AT = mmf for air gap + mmf for iron parts

$$= 3972 + 397 = 4369 \text{ A}$$

Current in exciting coil

$$I = \frac{AT}{T} = \frac{4369}{3000} = 1.456 \text{ A}$$

2.8 DESIGN OF FIELD (COILS) SYSTEM

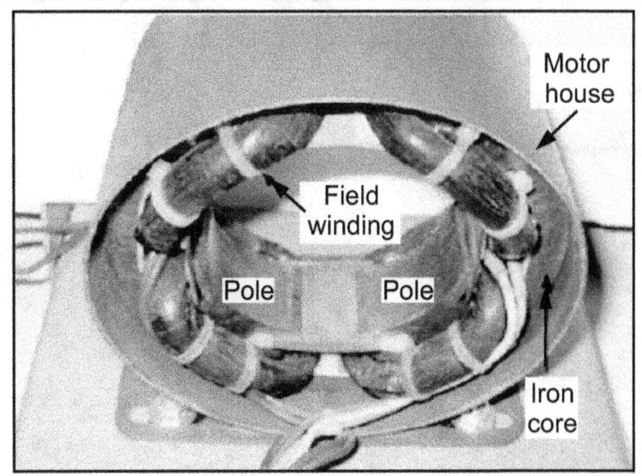

Fig. 2.7 : DC Motor stator with poles visible

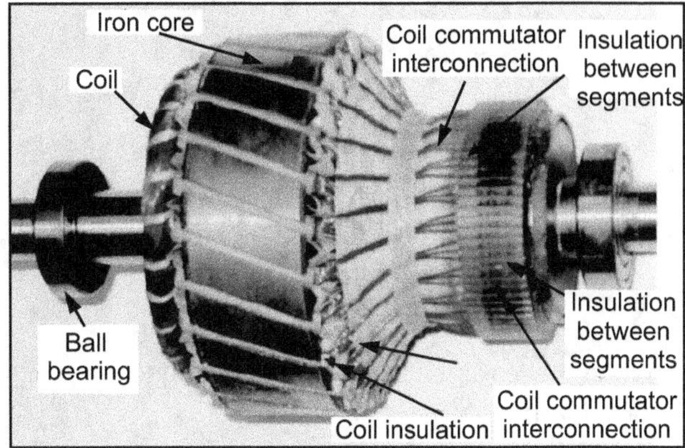

Fig. 2.8 : Rotor of DC motor

2.8.1 Magnetic Circuit

- The path of magnetic flux is called magnetic circuit.
- Magnetic circuit of dc machine comprises of yoke, poles, airgap, armature teeth and armature core.
- Flux produced by field coils emerges from N pole and cross the air gap to enter the armature tooth. Then it flows through armature core and again cross the air gap to enter the S pole.

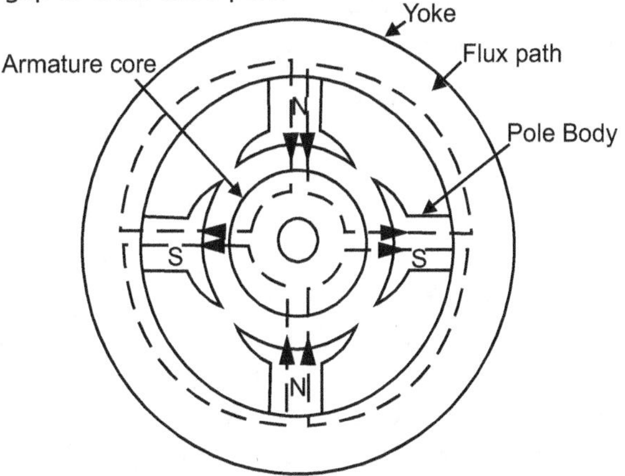

Fig. 2.9 : Magnetic circuit of 4-pole DC machine

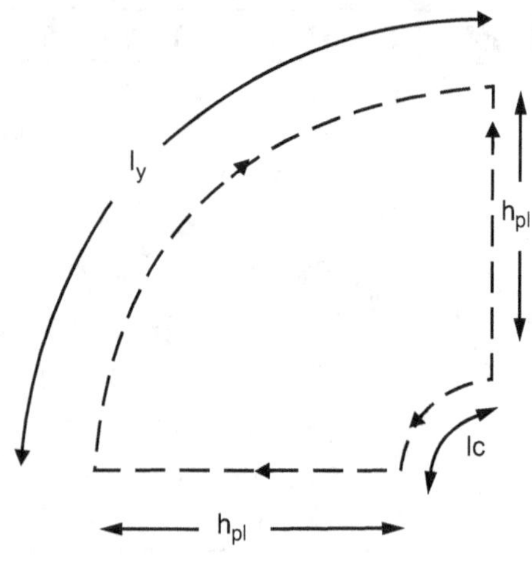

Fig. 2.10

Let B_g – Max. flux density in the core

         K_g – Gap contraction factor

         l_c – Length of magnetic path in the core

         l_y – Length of magnetic path in the yoke

         d_s – Depth of the slot

         d_c – Depth of core

         h_{pl} – Height of field pole

         D_m – Mean diameter of armature

- When the leakage flux is neglected magnetic circuit of a DC machine consists of following:
  - ➢ Yoke
  - ➢ Pole and pole shoe
  - ➢ Air gap
  - ➢ Armature teeth
  - ➢ Armature core

2.8.2 Total MMF to be Developed by each Pole

Total MMF to be developed by each pole is given by the sum of MMF required for the above five sections.

     MMF for air gap

$$AT_g = 800000\, B_g\, K_g\, l_g$$

     MMF for teeth

$$AT_t = at_t \times d_s$$

     MMF for core

$$AT_c = at_c \times l_c/2$$

     MMF for pole

$$AT_p = at_p \times h_{pl}$$

     MMF for yoke

$$AT_y = at_y \times l_y/2$$

at_t, at_c, at_p, at_y - are determined B-H curves

$$l_c = \pi D_m/P = \pi(D - 2d_s - d_c)/P$$

$$l_y = \pi D_{my}/P = \pi(D + 2l_g + 2h_{pl} + d_y)/P$$

$$AT\ total = AT_g + AT_t + AT_c + AT_p + AT_y$$

2.8.3 Design of Field System

Design of field system consists of poles, pole shoe and field winding.

Types:

1. Shunt field
2. Series field

Shunt Field Winding - have large no of turns made of thin conductors, because current carried by them is very low

Series field winding is designed to carry heavy current and so it is made of thick conductors/strips

Field coils are formed, insulated and fixed over the field poles.

2.8.4 Factors to be Considered in Design

- MMF/pole &flux density
- Losses dissipated from the surface of field coil
- Resistance of the field coil
- Current density in the field conductors

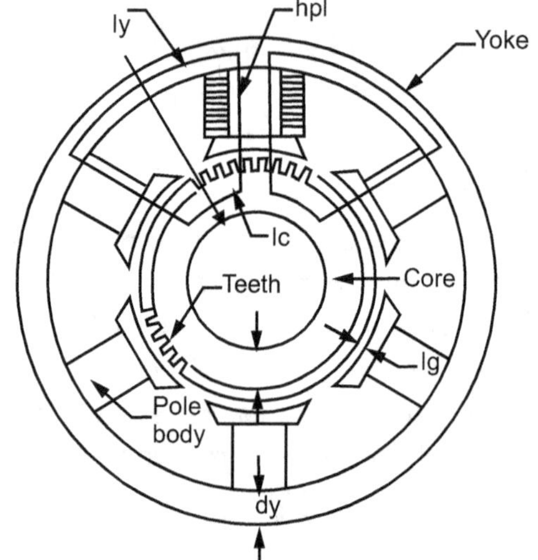

Fig. 2.11 : Magnetic circuit of dc machine

2.8.5 Tentative Design of Field Winding

     Let,

         AT_{fl} - MMF developed by field winding at full load

         Q_f - Copper loss in each field coil (W)

         q_f - Permissible loss per unit winding surface for normal temperature rise (W/m^2)

         S_f - Copper space factor

         ρ - Resistivity (Ω –m)

         h_f - Height of winding (m)

         d_f - Depth of winding (m)

         S - Cooling surface of field coil (m^2)

L_{mt} - Length of mean turn of field winding (m)

R_f - Resistance of each field coil (ohms)

T_f - Number of turns in each field coil

A_f - Area of each conductor of field winding (m^2)

I_f - Current in the field winding (A)

δ_f - Current density in the field winding (A/mm^2)

Cooling surface of the field winding,

$$S = 2L_{mt}h_f \qquad \ldots (2.50)$$

Permissible copper loss in each field coil,

$$S_{qf} = 2L_{mt}h_f q_f \qquad \ldots (2.51)$$

Area of X-section of field coil

$$= h_f d_f \qquad \ldots (2.52)$$

Area of copper in each section

$$= S_f h_f d_f \qquad \ldots (2.53)$$

i.e, $T_f a_f = S_f h_f d_f \qquad \ldots (2.54)$

Copper loss in each field coil,

$$Q_f = I_f^2$$
$$R_f = I_f^2 (T_f \, \rho L_{mt})/a_f$$
$$Q_f = I_f^2 \left(\frac{T_f \rho L_{mt}}{a_f}\right) = (a_f \delta_f)^2 \left(\frac{T_f \rho L_{mt}}{a_f}\right)$$
$$Q_f = \delta_f^2 \, \rho(T_f a_f \cdot L_{mt}) \qquad \ldots (2.55)$$
$$Q_f = \delta_f^2 \, \rho \times \text{Volume of Copper}$$

i.e., Copper loss $\alpha \, \delta_f^2$ (Square of the current density)

To have temperature rise within the limit, the copper loss should be equal to the permissible loss.

Using Eqns. (2.51) & (2.55),

$$2L_{mt} \, h_f \, q_f = \delta_f^2 \, \rho L_{mt} (S_f h_f d_f) => \delta_f = \sqrt{\frac{2q_f}{\rho S_f d_f}} \qquad \ldots (2.56)$$

MMF per metre height of field winding

$$= \frac{AT_{fl}}{h_f} = \frac{I_f T_f}{h_f}$$
$$= \frac{\delta_f a_f T_f}{h_f} = \frac{\delta_f S_f d_f h_f}{h_f}$$
$$= \delta_f S_f d_f = \sqrt{\frac{2q_f}{\rho S_f d_f}} \cdot S_f d_f$$
$$= \sqrt{\frac{2q_f S_f d_f}{\rho}} \qquad [\because \rho = 2 \times 10^{-8} \, \Omega.m\,]$$

MMF per meter height $= 10^4 \times \sqrt{q_f S_f d_f} \qquad \ldots (2.57)$

Normal values:

Permissible loss, q_f -700W/m^2

Copper Space factor, S_f :

Small wires: 0.4

Large round wires: 0.65

Large rectangular conductors: 0.75

Depth of the field winding, d_f :

Armature Dia. (m)	Winding Depth (mm)
0.2	30
0.35	35
0.5	40
0.65	45
1.00	50
1.00 and above	55

Height of field,

Total height of the pole,

$$h_f = \frac{AT_{fl}}{\text{Armature Turns per meter height}}$$

Using Eq. (2.57),

$$h_f = \frac{AT_{fl} \times 10^{-4}}{\sqrt{q_f S_f d_f}}$$

$$h_{pl} = h_f + h_s + \text{height for insulation and curvature of yoke}$$

where,

h_s - Height of the pole shoe (≈ 0.1 to 0.2 of the pole height)

2.8.6 Design of Shunt Field Winding

- Involves the determination of the following information regarding the pole and shunt field winding
  - Dimensions of the main field pole,
  - Dimensions of the field coil,
  - Current in shunt field winding,
  - Resistance of coil,
  - Dimensions of field conductor,
  - Number of turns in the field coil,
  - Losses in field coil.

- **Dimensions of the Main Field Pole:** For rectangular field poles
  - Cross sectional area, length, width, height of the body

- For cylindrical pole
  - Cross sectional area, diameter, height of the body

Area of the pole body can be estimated from the knowledge of flux per pole, leakage coefficient and flux density in the pole

Leakage coefficient (C_l) depends on power output of the DC machine

Bp in the pole 1.2 to 1.7 wb/m^2

$$\phi_p = C_l \cdot \phi$$

$$A_p = \phi_p / B_p$$

When circular poles are employed, C.S.A will be a circle

$$A_p = \pi d_p^2 / 4$$

$$d_p = \sqrt{4A_p / \pi}$$

When rectangular poles employed, length of pole is chosen as 10 to15 mm less than the length of armature

$$L_p = L - (0.001 \text{ to } 0.015)$$

Net iron length

$$L_{pi} = 0.9 \, L_p$$

Width of pole,

$$b_p = A_p / L_{pi}$$

Height of pole body

$$h_p = h_f + \text{thickness of insulation and clearance}$$

Total height of the pole

$$h_{pl} = h_p + h_s$$

- Field coils are former wound and placed on the poles
- They may be of rectangular or circular cross section depends on the type of poles
- Dimensions – L_{mt}, depth, height, diameter

　　Depth(d_f) – depends on armature

　　Height (h_f) – depends on surface required for cooling the coil and no. of turns (T_f)

　　h_f, T_f – cannot be independently designed

　　L_{mt} – Calculated using the dimensions of pole and depth of the coil

- For rectangular coils

$$L_{mt} = 2(L_p + b_p + 2d_f) \text{ or } (L_o + L_i)/2$$

Where L_o – length of outer most turn & L_i – length of inner most turn

- For cylindrical coils

$$L_{mt} = \pi(d_p + d_f)$$

- **No of Turns in Field Coil:** When the ampere turns to be developed by the field coil is known, the turns can be estimated

Field ampere turns on load,

$$AT_{fl} = I_f \cdot T_f$$

Turns in field coil,

$$T_f = AT_{fl} / I_f$$

1. Power Loss in the Field Coil:

- Power loss in the field coil is copper loss, depends on Resistance and current

- Heat is developed in the field coil due to this loss and it is dissipated through the surface of the coil

- In field coil design, loss dissipated per unit surface area is specified and from which the required surface area can be estimated.

- Surface area of field coil – depends on L_{mt}, depth and height of the coil

　　L_{mt} – estimated from dimensions of pole

　　Depth – assumed (depends on diameter of armature)

　　Height – estimated in order to provide required surface area

- Heat can be dissipated from all the four sides of a coil. i.e., inner, outer, top and bottom surface of the coil

Inner surface area = $L_{mt} (h_f - d_f)$

Outer surface area = $L_{mt} (h_f + d_f)$

Top and bottom surface area = $L_{mt} \, d_f$

Total surface area of field coil, $S = L_{mt} (h_f - d_f) +$
$= L_{mt} (h_f + d_f) + L_{mt} \, d_f + L_{mt} \, d_f$

$$S = 2L_{mt} \, h_f + L_{mt} \, d_f = 2L_{mt} (h_f + d_f)$$

- Permissible copper loss,

$$Q_f = S.q_f \; [q_f \text{ -Loss dissipated/ unit area}]$$

- Substitute S in Q_f,

$$Q_f = 2L_{mt} (h_f + d_f).q_f$$

- Actual Cu loss in field coil $= I_f^2 R_f = E_f^2 / R_f$

Substituting

$$R_f = (\rho L_{mt} \, T_f) / a_f$$

- Actual Cu loss in field coil $= E_f^2 \cdot a_f / (\rho L_{mt} \, T_f)$

$$2L_{mt} q_f (h_f + d_f) = \frac{E_f^2 a_f}{\rho L_{mt} \, T_f}$$

- Conductor area in field coil

$$= \text{No. of turns} \times \frac{\text{area of X-section}}{\text{of field conductor}}$$

$$= T_f a_f$$

- Conductor area in field coil

$$= \text{copper space factor} \times \frac{\text{area of X-section}}{\text{of field conductor}}$$

2. Procedure for Design of Shunt Field Winding

Step 1 : Determine the dimensions of the pole. Assume a suitable value of leakage coefficient and B = 1.2 to 1.7 T

$$\phi_p = C_l . \phi$$

$$A_p = \phi_p/B_p$$

When circular poles are employed, C.S.A will be a circle

$$A_p = \pi d_p^2/4 : d_p = \sqrt{(4A_p/\pi)}$$

When rectangular poles employed, length of pole is chosen as 10 to 15 mm less than the length of armature

$$L_p = L - (0.001 \text{ to } 0.015)$$

Net iron length

$$L_{pi} = 0.9 \, L_p$$

Width of pole $= A_p/L_{pi}$

Step 2 : Determine L_{mt} of field coil

Assume suitable depth of field winding

For rectangular coils

$$L_{mt} = 2(L_p + b_p + 2d_f) \text{ or } (L_o + L_i)/2$$

For cylindrical coils $L_{mt} = \pi(d_p + d_f)$

Step 3 : Calculate the voltage across each shunt field coil

$$E_f = (0.8 \text{ to } 0.85) \, V/P$$

Step 4 : Calculate C.S.A of filed conductor

$$A_f = \rho L_{mt} \, AT_{fl}/E_f$$

Step 5 : Calculate diameter of field conductor

$$d_{fc} = \sqrt{(4a_f/\pi)}$$

Diameter including thickness $d_{fci} = d_{fc} +$ insulation thickness Copper space factor $S_f = 0.75(d_{fc}/d_{fci})^2$

Step 6 : Determine no. of turns (Tf) and height of coil (hf) They can be determined by solving the following two equations

$$2L_{mt}(h_f + d_f) = E_f^2 \, a_f/\rho L_{mt} \, T_f$$

$$T_f.a_f = S_f.h_f.d_f$$

Step 7 : Calculate R_f and I_f :

$$R_f = T_f. \, \rho L_{mt} \, /a_f$$

$$I_f = E_f/R_f$$

Step 8 : Check for δ_f

$$\delta_f = I_f \, / \, a_f$$

$$\delta_f - \text{ not to exceed } 3.5A/mm^2.$$

If it exceeds then increase a_f by 5% and then proceed again

Step 9 : Check for desired value of AT

$$AT_{actual} = I_f T_f$$

$$AT_{desired} = 1.1 \text{ to } 1.25 \text{ times armature MMF at full load}$$

When AT_{actual} exceeds the desired value then increase the depth of field winding by 5% and proceed again.

Check for Temp Rise:

Actual copper loss $= I_f^2 \, R_f$

Surface area $= S = 2L_{mt} \, (h_f + d_f)$

Cooling coefficient $C = (0.14 \text{ to } 0.16)/(1 + 0.1 \, V_a)$

$$\theta_m = \text{Actual copper loss X } (C/S)$$

If temperature rise exceeds the limit, then increase the depth of field winding by 5% and proceed again.

2.8.7 Design of Series Field Winding

Step 1: Estimate the AT to be developed by series field coil,

$$AT \, /pole = (I_z \cdot (Z/2))/P$$

For compound m/c,

$$AT_{se} = (0.15 \text{ to } .25) \, (I_z \cdot Z)/2P$$

For series m/c, $AT_{se} = (1.15 \text{ to } 1.25) \, (I_z \cdot Z)/2P$

Step 2: Calculate the no. of turns in the series field coil,

$$T_{se} = AT_{se}/I_{se} \text{ (Corrected to an integer)}$$

Step 3: Determine cross sectional area of series field conductor,

$$a_{se} = I_{se} \, /\delta_{se}$$

Normally, $\delta_{se} = 2 \text{ to } 2.3 \text{ A } /mm^2$

Step 4 : Estimate the dimension of the field coil

Conductor area of field coil $= T_{se}.a_{se}$

Also Conductor area of field coil $= S_{fse}.h_{se}.d_{se}$

When circular conductors are used

$$S_{fse} = 0.6 \text{ to } 0.7$$

For rectangular conductors, S_{fse} – depends on thickness and type of insulation

On equating above two expressions,

$$T_{se}.a_{se} = S_{fse}.h_{se}.d_{se}$$

$$h_{se} = (T_{se}.a_{se} \,)/(S_{fse}.d_{se})$$

EXERCISE

1. Describe a design procedure for a round wire heating element

2. Develop the design procedure of choke.

3. Derive the calculation of resistence steps of motor starters.

4. How would you calculate the size of wire of field rheostats for generators and motors.

5. Develop the design procedure for shunt field winding

6. Describe a design procedure for series field winding.

7. Discuss a design of lifting magnet coil.

3.1 INTRODUCTION

- Armature winding of a machine is defined as an arrangement of conductors designed to produce emfs by relative motion in a heteropolar magnetic field. The action of rotating electromagnetic machines is dependent upon the conversion of power that takes place through the medium of magnetic field : an emf being induced in the winding that experiences a change of flux linkages.

- Electrical machines employ groups of conductors distributed in slots over the periphery of the armature. The groups of conductors are connected in various types of series – parallel combinations to form armature winding. The conductors are connected in series so as to increase the voltage rating while they are connected in parallel to increase the current rating. Some of the commonly used terms associated with windings are explained below :

1. **Conductor :** The active length of wire or strip in the slot i.e. length of wire responsible for electromagnetic energy conversion.

2. **Turn :** A turn consists of two conductors separated from each other by a pole pitch or nearly so, and connected in series as shown in Fig. 3.1 (a). The conductors forming a turn are kept a pole pitch apart in order that the emfs in two are additive to produce maximum resultant emf.

3. **Coil :** A coil may consist of a single turn or may consist of many turns, placed in almost similar magnetic position, connected in series. In the former case, the coil is called a single turn coil while in the latter case, it is known as multi-turn coil. A single turn coil is shown in Fig. 3.1 (a). Fig. 3.1 (b) shows a 3-turn coil. However, any number of turns could be included in a coil before bringing out the coil ends.

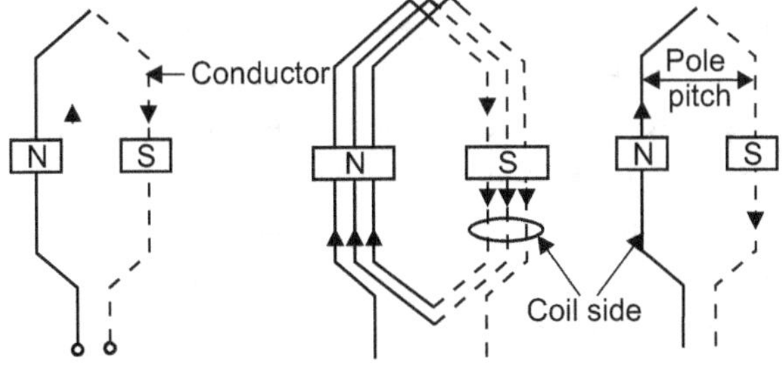

(a) Single coil   **(b) 3-turn coil**   **(c) Multi turn coil**

Fig. 3.1 : Different types of winding coils representations

4. **Coil Side :** A coil consists of two coil sides which are placed in two different slots which are almost a pole pitch apart. The group of conductors on one side of the coil form one coil side while the conductors on the other side of coil situated a pole pitch (or approximately a pole pitch apart) from the second coil side. Fig. 3.1 (c) shows the representation of a coil irrespective of whether it is a single turn or a multi turn coil.

The connections joining the conductors form the end connectors or in the mass, the overhang or end-winding. When the two coil sides forming a coil are spaced exactly one pole pitch apart they are said to be of full pitch. However, the coil span may be less than a pole pitch, in which case the coil is described as short pitched or chorded.

5. **Coil Group :** One or more coil single coils formed in a group forms the coil group.

6. **Winding :** Number of coils arranged in coil group is said to be a winding.

7. **Pole Pitch:** Distance between the poles in terms of slots is called pole pitch.

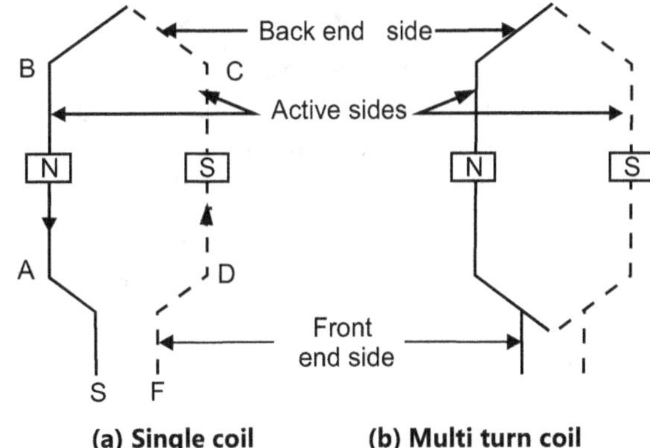

(a) Single coil   **(b) Multi turn coil**

Fig. 3.2 : Single and Multi turn coils

8. **Full Pitch Winding:** If the coil pitch for a winding is equal to pole pitch the winding is called full pitch winding Fig. 3.3

9. **Chorded Winding:** When the pitch of the winding is less than the full pitch or pole pitch then the winding is called short pitch winding or chorded winding. Fig. 3.3.

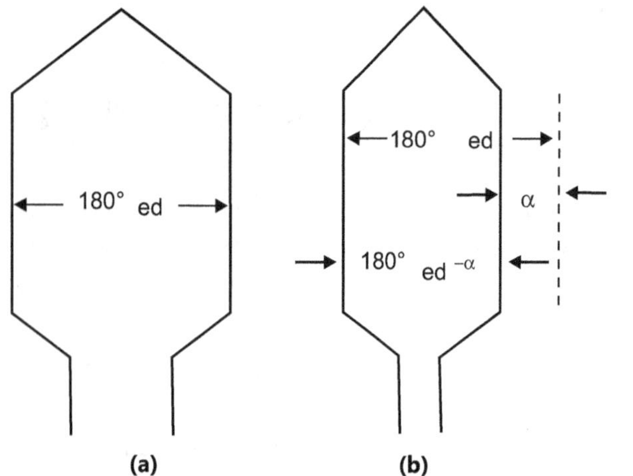

Fig. 3.3 : Full pitched and short pitched coils

3.2 SINGLE LAYER AND TWO (DOUBLE) LAYER WINDING

There are two basic physical types for the windings. They deal differently with the mechanical problem of arranging coils in sequence around the armature. The two types are :

1. Single layer winding, and

2. Double layer winding.

1. **Single Layer Winding :** Fig. 3.4 (a) shows an arrangement for a single layer winding. In this type of winding arrangement, the whole of the slot is occupied by one coil side if a coil. Single layer windings are not used for machines having commentators. However, some advantages of these windings are :

* Single layer windings allow the use of semi-enclosed and closed types of slots.

* With single layer windings, coils preformed at one end can be used. These coils are pushed through the slots from one end of the core and are connected up during the winding process at the other end which happens to be free end. The use of preformed coils wherein the insulation can be properly applied and consolidated is a great advantage especially in large output high voltage machines.

* High voltage single layer windings use small groups of concentrically placed coils. These coils are interlinked in such a way as to minimize both space taken up outside the slot and in the overhang connections. A coil group employs coils of different coil spans (and hence if different shapes), with average span of coil group being equal to pole pitch. Small rating low voltage induction machines use mush type single layer winding.

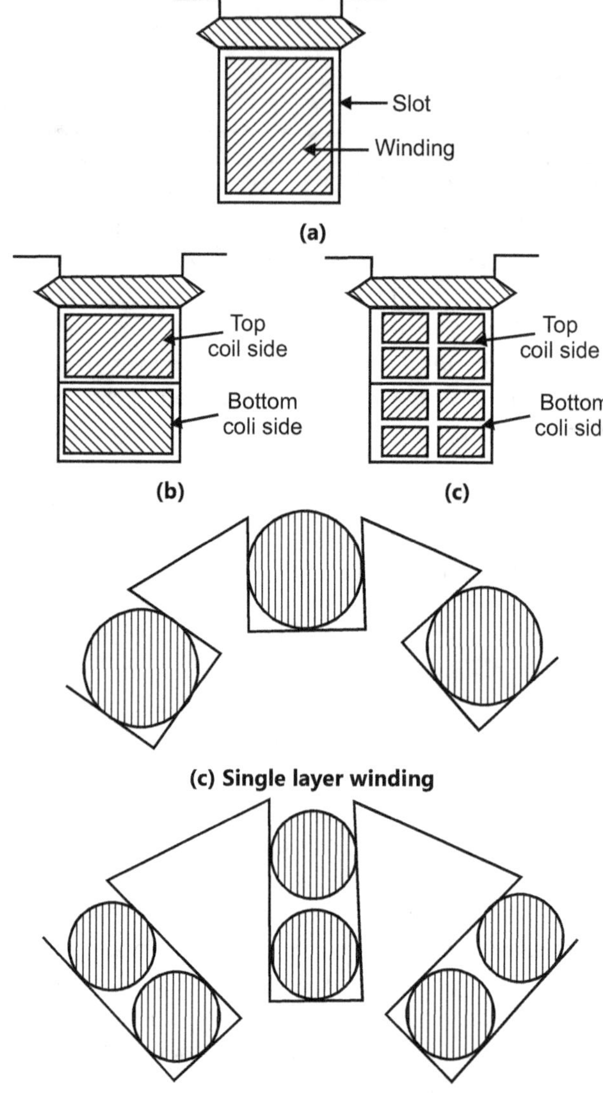

Fig. 3.4 : Single and double layer windings

2. **Double Layer Windings :** The two layer windings have identical coils with one coil side of each coil laying in top half of the slot and the other coil side in bottom half of another slot exactly or approximately one pole pitch away. Fig. 3.4 (b) shows a double layer winding with 2 coil sides per slot. Each layer may contain more than one coil side in case large number of coils are required. Fig. 3.4 (c) shows an arrangement wherein there are 8 coil sides per slot. Open slots are frequently used to house double layer windings.

The advantages of double layer windings are :

* Double layer windings give a neat arrangement with all coils being identical.

* Double layer windings give greater flexibility in design because of the case with which the coil span (or coil pitch) can be chosen.

3.3 CLASSIFICATION OF WINDINGS

Armature windings are classified in two categories

1. Closed type and
2. Open type

1. Closed Windings :

- In this winding there is a closed path around the armature. Thus if starting at any point, the winding is followed through all its turns, the starting point is reached again. The current flowing into a closed winding is through brushes placed on a commutator whose segments are connected to different armature coils. The armature current divides itself into parallel paths which retain their spatial orientation irrespective of the position of the rotating winding.

- Even though the coils carrying this current are continuously changing, the view of the winding from the brush always remains the same and thus the brushes maintain the same polarity. This is possible only through the use of a commutator. The closed windings are used in d.c. and a.c. commutator machines because in these machines they are a must in order that the commutator may accomplish its purpose as a switching device. Closed windings are always double layer type.

2. Open Windings :

- In alternating current machines where commutator is not used, it is not necessary to use a closed windings. These machines use open winding. In open windings, the armature may be left open at one or more points.

- The ends of each section of the winding can be brought out to terminals and then any desired inter-connection can be made externally. Where an open winding is possible, it is universally preferred to a closed winding because of the flexibility in design and the freedom of connections.

- Open windings may be wither single layer or double layer type and are used in a.c. machines like induction and synchronous machines.

(a)

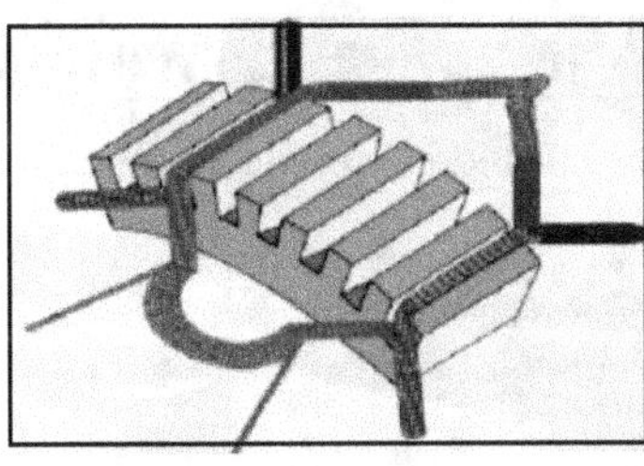

(b)

Fig. 3.5 : Photographs of the windings and coils

- Closed winding which are adapted for connection to a commutator are usually called D.C. Armature Windings while open windings which are exclusively used for a. c. machines are known as A.C. Armature Windings. These two types of windings are discussed in details in

D.C. ARMATURE WINDINGS

3.4 COILS AND COIL SIDES

- D. C. armature windings are closed and double layer type with each coil, consisting of an upper coil side at the top of one slot and a lower coil side situated at the bottom of another slot approximately a pole pitch away. The coils are diamond-shaped made in special forming machines. The back and front bends are constructed so that one coil side is on a circumferentially higher level than the other. This construction makes it possible to place the higher coil side in the top of a slot, and the lower coil side in the bottom of another slot (Fig. 3.6).

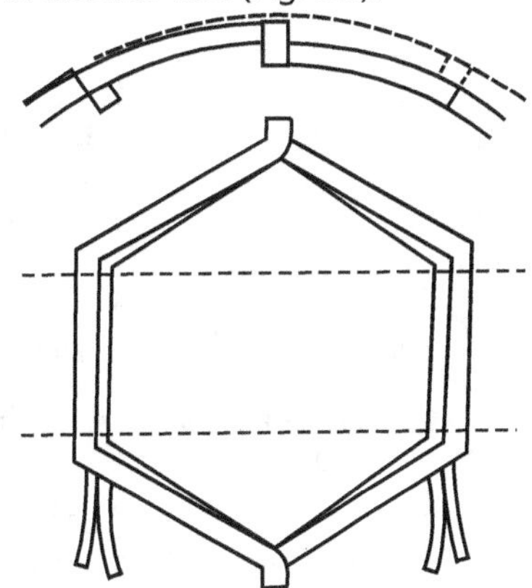

Fig. 3.6 : Diamond-shaped coil

- The number of slots required in a machine is equal to the number of coils if the slot contains two coil sides, one in each layer. But in high voltage and low speed machines, the number of coils is large and it therefore, becomes, necessary to have 4, 6, 8 or even more coil sides per slot (2, 3, 4 or more coil sides per layer) as it is not always possible to increase the number of slots. Fig. 3.7 shows a winding having 6 coil sides per slot.

3.5 TYPES AND SHAPES OF WINDING WIRES

The winding wires used in electrical motors are classified as follows.

1. Round wires

2. Rectangular straps

3. Stranded wires

1. **Round Wires:** It has thin and thick conductors and are used in semi-closed slot type motors and mush winding rotors. It is wounded in reels and available in Kilograms.

2. **Rectangular Straps:** It is used in open type slot motors. These conductors are available as long straps in meters. They are used in the following places. 1) Low voltage motor windings. 2) Used as conductor in high current motor. 3) Series field motor winding coils.

3.6 NUMBERING SCHEME

All the top civil sides are numbered odd while all the bottom coil sides are numbered even as shown in Fig. 3.7 It is to be noted that coil side 2 is under coil side 1, while coil side 4 is under coil side 3. The numbers 1, 3, 5, 7 etc. at the top and 2, 4, 6, 8 etc. at the bottom progress clockwise.

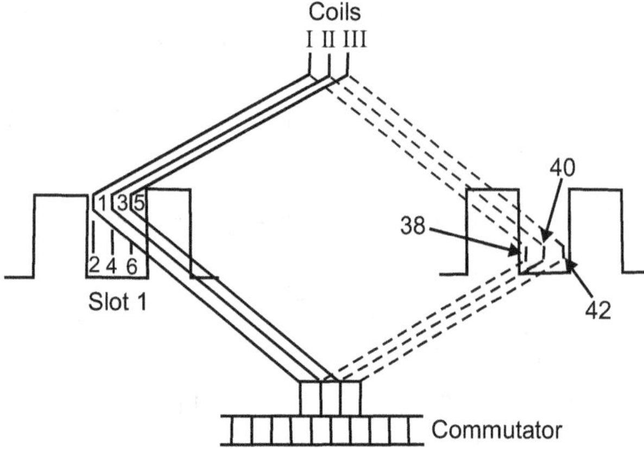

Fig. 3.7 : Winding with 6 coil sides per slot

3.7 COIL SPAN

*   The number of teeth embraced by the two coil sides of a coil is known as its coil span. It is usual that the coil span of each armature coil be made equal to the pole pitch (in terms of number of teeth embraced) although the coils are deliberately chorded in special cases, while with certain winding arrangements, it is not possible to have the coil span exactly equal to the pole pitch i.e. where the number of slots is not divisible by the number of poles.

*   Consider a winding with 37 slots and 4 poles; a pole pitch covers $\frac{37}{4} = 9\frac{1}{4}$ slots. Thus the coil span can either by 9 or 10 slots (or teeth). It is usual to take the nearest whole number, 9 in this case. If the coil span is chosen as 9 slots and if the top coil side of a coil lies in slot 1, the bottom coil side will lie in $(1 + 9) = 10^{th}$ slot.

3.8 TYPES OF D. C. WINDING

Modern d.c. machines employ two general types of windings :

1. Lap windings

2. Wave windings.

These two types of windings differ in two ways

(i)   Number of circuits between positive and negative brushes,

(ii) The manner in which the coil ends are connected. However the coils of both lap and wave windings are identically formed

3.8.1 Lap Winding

*   The most commonly used windings are the simplex lap and wave windings. In the simplex lap winding (Fig. 3.8), the finish F_1 of coil I is connected to start S_2 of coil II starting under the same pole as start S_1 of coil I. The connector connecting F_1 and S_2 is connected to commutator segment 2 adjacent to segment 1 to which start S_1 of the coil I and the finish of last coil are connected.

*   The winding in which successive coils overlap each other hence it is called lap winding. In this winding end of one coil is connected to the commutator segment and start of the adjacent coil situated under the same pole as shown in Fig. 3.8. Lap winding is further divided as simplex and Duplex lap winding.

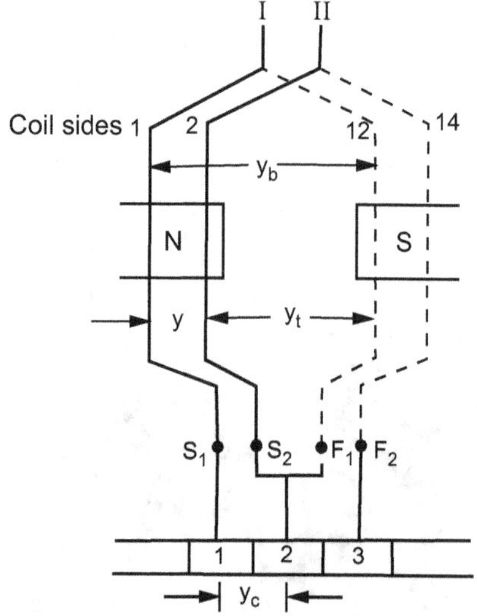

Fig. 3.8 : Lap winding

3.8.2 Simplex Lap Winding

- In this type of winding finish F_1 of the coil 1 is connected to the start S_2 of coil 2 starting under the same pole as start S_1 of coil 1.

- We have back pitch $y_b = 2c/p \pm k$ where c = number of coils in the armature, p = number of poles, k = an integer to make y_b an odd integer.

Important Rules for Lap Winding:

Let Z = Number of conductors

          P = number of poles

          Y_b = Back pitch

          Y_f = Front pitch

          Y_c = Commutator pitch

          Y_a = Average pole pitch

          Y_p = Pole pitch

          Y_R = Resultant pitch

- Y_b (Back pitch) and Y_f (Front pitch) must be approximately equal to Y_p (Pole pitch)

- Y_b (Back pitch) must be less or greater than Y_f (Front pitch) by 2m where m is the multiplicity of the winding. When Y_b is greater than Y_f the winding progresses from left to right and is known as progressive winding. When Y_b is lesser than Y_f the winding progresses from right to left and is known as retrogressive winding. Hence $Y_b = Y_f \pm 2m$.

- Y_b and Y_f must be odd.

- Y_b and Y_f may be equal or differ by $\pm$ 2. + for progressive winding, - for retrogressive winding

- $Y_a = (Y_b + Y_f) / 2 = Y_p$

- Y_R (Resultant pitch) is always even.

- Y_c = m, m = 1 for simplex winding; m = 2 for duplex winding

- Number of parallel paths = mp = number of brushes.

3.8.3 Wave Winding

- In wave winding the end of one coil is not connected to the beginning of the same coil but is connected to the beginning of another coil of the same polarity as that of the first coil as shown in Fig. 3.9.

3.8.4 Simplex Wave Winding

In the simples wave winding (Fig. 3.9), the finish F_1 of coil 1 is connected to start S_{11} of coil XI (for this particular case) which is located under a similar pole to one under which coil I started. This means that the start of two consecutive coils (S_1 and S_{11}) are two pole pitches apart in wave winding.

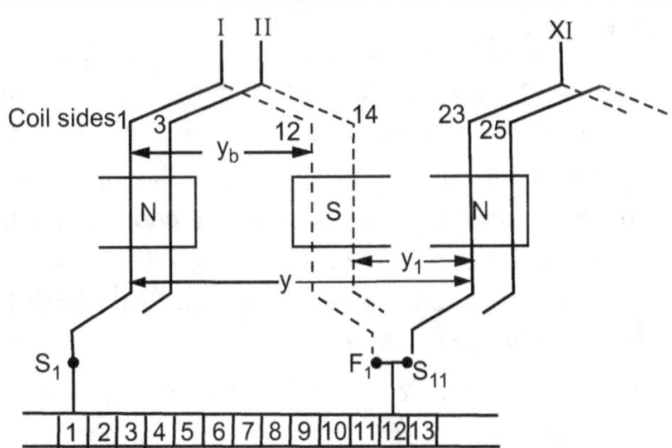

Fig. 3.9 : Wave winding

Important Rules for Wave Winding:

- Y_b (Back pitch) and Y_f (Front pitch) must be approximately equal to Y_p (Pole pitch)

- Y_b and Y_f must be odd.

- Y_b and Y_f may be equal or differ by ± 2. + for progressive winding, - for retrogressive winding

- $Y_c = (Y_b + Y_f) / 2$ and should be a whole number.

3.8.5 Dummy Coils

- The wave winding is possible only with particular number of conductors and poles and slots combinations. Sometimes the standard stampings do not consist of the number of slots according to the design requirements and hence the slots and conductor combination will not produce a mechanically balanced winding.

- Under such conditions some coils are placed in the slots, not connected to the remaining part of the winding but only for mechanical balance. Such windings are called dummy coils.

3.9 WINDING PITCHS

3.9.1 Back Pitch

- The distance between top and bottom coil sides of a coil, measured around the back of the armature (nway from the commutator) is called the back pitch and is designated as v_h. The back pitch is measured in terms of coil sides.

- Since y_h is the difference between an even and an odd number and must, therefore, always be an odd number. In Fig. 3.8 and Fig. 3.9, the two coil sides of coil No. 1 are numbered 1 and 12, therefore $y_b = 12 - 1$ = 11 in these cases. The back pitch of a coil determine the size of the coil and is nearly equal to coil rides per pole.

3.9.2 Front Pitch

- The distance between two coil connected to the same commutator segment is called the front pitch and is designated as y_f. The front pitch for the lap winding shown in Fig. 3.8 is $y_f = (12 - 3) = 9$ and that for the wave winding shown in Fig. 3.9 is $y_f = (23 - 12) = 11$. It as noted from above that the front pitch for both lap and wave windings is odd.

- If we trace the connection for lap winding the movement is in the forward direction (coil side No. 12 to No. 23). Thus, the front pitch determines the type of winding only and it does not affect the size of the coils.

3.9.3 Winding Pitch

- The distance between the starts of two consecutive coils measured in terms of coil sides is called winding pitch and is designated as Y.

$$Y = y_b - y_f \text{ for lap winding} \qquad \text{... (3.1)}$$
$$Y = y_b + y_f \text{ for wave winding} \qquad \text{... (3.2)}$$

- For the simplex lap winding shown in Fig. 3.8; the winding pitch, $Y = 11 - 9 = 2$.

- While the back pitch for the simplex wave winding shown in Fig. 3.9 is

$$Y = 11 + 11 = 22$$

3.9.4 Commutator Pitch

- The distance between the two commutator segments to which the two ends (starts and finish) of a coil are connected is called the commutator pitch. It is measured in terms of commutator segments and is designated as y_c. Thus from Figs. 3.8 and 3.9,

$$y_c = 2 - 1 = 1 \text{ in the case of simplex lap}$$
$$\text{winding foe all cases ; and}$$
$$y_c = 12 - 1 = 11 \text{ in the case of simplex}$$
$$\text{wave winding of Fig. 3.9.}$$

- On tracing through the simplex lap winding in Fig. 3.8, a segment adjacent to the first one is reached after traversing through one coil, while for simplex wave winding (Fig. 3.9), the adjacent segment is reached after traversing through p/2 coils.

3.10 EQUALIZER RINGS OR EQUALIZER CONNECTIONS IN LAP WINDING

- This is the thick copper conductor connecting the equipotential points of lap winding for equalizing the potential of different parallel paths.

3.11 SEQUENCE DIAGRAM OR RING DIAGRAM

- The diagram obtained by connecting the conductors together with their respective numbers. This diagram is used for finding the direction of induced emf and the position of brushes.

3.12 DEVELOPED DIAGRAM

- Instead of dealing with circular disposition of the slots and the commutator segments, it is always advantageous to work with the developed diagram of the armature slots and the commutator segments (sequence diagram) as elaborated in Fig. 3.10

- In the Fig. 3.10 actual armature with 8 slots and 8 commutator segments are shown.

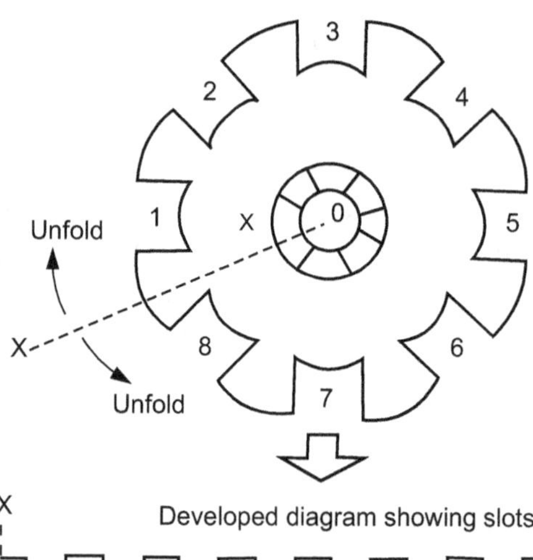

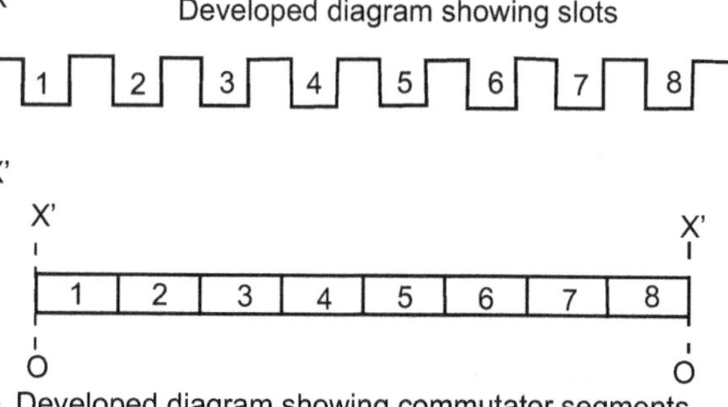

Developed diagram showing commutator segments

Fig. 3.10: Sequence diagram

3.13 POLE PITCH, POLE ARC AND POLE FIXING ON A DEVELOPED WINDING DIAGRAM

- The distance corresponding to a pole is called the **Pole Pitch (τ)**.

- The actual distance covered by the pole or pole shoe is called the **Pole arc**.

- (Generally the pole arc lies between 60 and 70 percent of the pole pitch)

- Let the pole arc be **70 %** of the pole pitch.
- The circumference of the armature is **πd**.
- Where d is the diameter of the armature or 360° Mech.
- However the distance between the first and last conductor + the distance between the 2 conductors considered to draw the developed winding diagram can be taken as the circumference of the armature

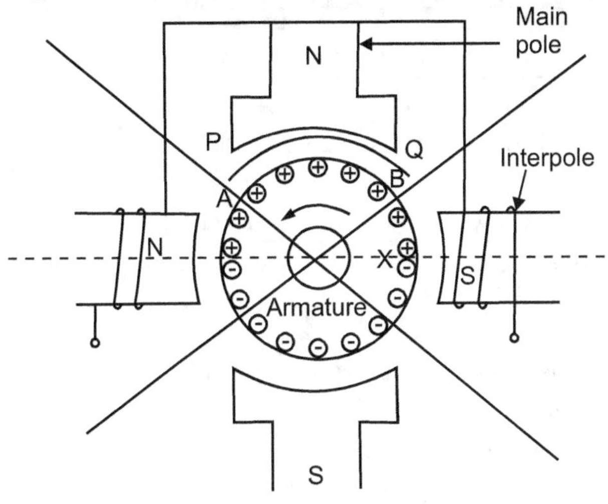

Fig. 3.11: DC Machine

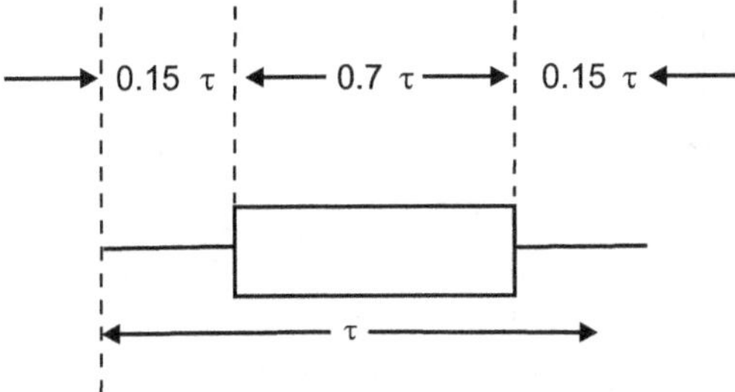

Fig. 3.12: Pole Pitch on a developed Winding Diagram

3.13.1 Brush Width

Minimum brush width	=	one segment width for simplex lap or wave winding
Minimum brush width	=	two segment width for duplex lap or wave winding
Minimum brush width	=	three segment width for triplex lap or wave winding

SOLVED EXAMPLES

Example 3.1 : *Develop a single layer lap winding diagram for a DC machine having 32 armature conductors and 4 poles. Mark the poles, draw the sequence diagram, indicate the position of the brushes and show the direction of induced emf.*

Solution : Analyse the Problem : This particular problem states us to develop a single layer lap winding diagram for a DC machine having 32 armature conductors and 4 poles.

Drawing sequence diagram helps us to analyse the flow of current and also helps to locate the brushes.

Given Data :

Number of poles (P) = 4

In single layer, number of conductors (Z) = slots = 32

Step 1 : First calculate the pole pitch, front pitch Y_f and back pitch Y_b.

Hence, pole pitch = Z/P = 32/4 = 8

But pole pitch = $(Y_b + Y_f)/2$

Hence, $(Y_b + Y_f) = 8$

$(Y_b + Y_f) = 16$

But $Y_b = Y_f \pm 2m$

(where, m = 1 for simplex, 2 for duplex and 3 for triplex)

$Y_b = 9$ and $Y_f = 7$

Step 2 : Now, develop the winding table so as to establish the coil connection in various slots as shown in Table 3.1.

Table 3.1 : Winding Table

At the Back $Y_b = 9$	At the Front $Y_f = 7$	At the Back $Y_b = 9$	At the Front $Y_f = 7$
1 + 9 = 10	10 − 7 = 3	17 + 9 = 26	26 − 7 = 19
3 + 9 = 12	12 − 7 = 5	19 + 9 = 28	28 − 7 = 21
5 + 9 = 14	14 − 7 = 7	21 + 9 = 30	30 − 7 = 23
7 + 9 = 16	16 − 7 = 9	23 + 9 = 32	32 − 7 = 25
9 + 9 = 18	18 − 7 = 11	25 + 9 = 34(2)	34 − 7 = 27
11 + 9 = 20	20 − 7 = 13	27 + 9 = 36(4)	36 − 7 = 29
13 + 9 = 22	22 − 7 = 15	29 + 9 = 38(6)	38 − 7 = 31
15 + 9 = 24	24 − 7 = 17	31 + 9 = 40(8)	40 − 7 = 33(1)

(a)

(b)

(c)

Fig. 3.13 : Winding diagram and Ring diagram

Example 3.2 : *A DC machine as 18 armature slots with 2 conductors per slot and develop a lap winding diagram for the DC machine, mark the poles, draw the sequence diagram indicate the position of the brushes and show the direction of induced emf.*

Solution :

Analysing the problem : This particular problem states us to develop a double layer lap diagram for a DC machine having 36 armature conductors and 6 poles. Drawing sequence diagram helps us to analyse the flow of current and helps to locate the brushes.

Solution : Given data :

Number of poles (P) = 6

Number of armature slots = 18

In double layer number of conductors (Z) = 2 × number of armature slots

$$Z = 2 \times 18 = 36$$

Step 1 : First calculate the pole pitch, front pitch Y_f and back pitch Y_b.

Hence,	pole pitch	$= Z/P = 36/6 = 6$
But	pole pitch	$= (Y_b + Y_f)/2$
Hence,	$(Y_b + Y_f)/2$	$= 6$
	$(Y_b + Y_f)$	$= 12$
But	Y_b	$= Y_f \pm 2m$

(where m = 1 for simplex, 2 for duplex and 3 for triplex)

$$(Y_b - Y_f) = 2$$

Solving Eq. we get,

$$Y_b = 7 \text{ and } Y_f = 5$$

Step 2 : Now, develop the winding table so as to establish the coil connection in various slots as shown in Table 3.2.

Table 3.2 : Winding Table

Sr. No.	At the Back $Y_b = 7$	At the Front $Y_f = 5$	Sr. No.	At the Back $Y_b = 7$	At the Front $Y_f = 5$
1	1 + 7 = 8	8 – 5 = 3	10	19 + 7 = 26	26 – 5 = 21
2	3 + 7 = 10	10 – 5 = 5	11	21 + 7 = 28	28 – 5 = 23
3	5 + 7 = 12	12 – 5 = 7	12	23 + 7 = 30	30 – 5 = 25
4	7 + 7 = 14	14 – 5 = 9	13	25 + 7 = 32	32 – 5 = 27
5	9 + 7 = 16	16 – 5 = 11	14	27 + 7 = 34	34 – 5 = 29
6	11 + 7 = 18	18 – 5 = 13	15	29 + 7 = 36	36 – 5 = 31
7	13 + 7 = 20	20 – 5 = 15	16	31 + 7 = 38(2)	(2)38 – 5 = 33
8	15 + 7 = 22	22 – 5 = 17	17	33 + 7 = 40(4)	(4)40 – 5 = 35
9	17 + 7 = 24	24 – 5 = 19	18	35 + 7 = 42(6)	(6)42 – 5 = 37(1)

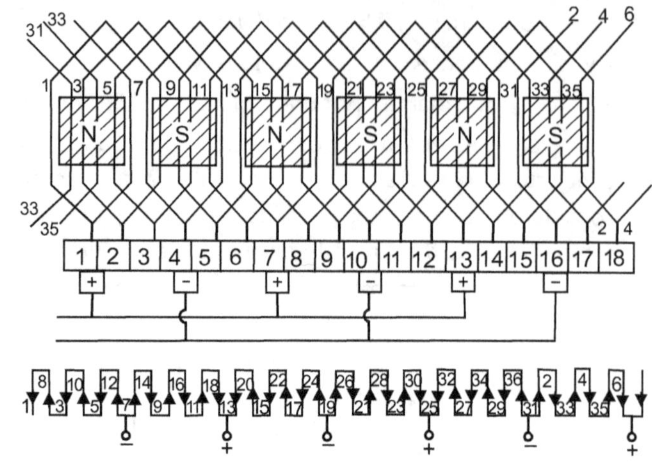

Fig. 3.14 : Winding diagram and Ring Diagram

Example 3.3 : *A DC machine as 28 armature conductors and 4 poles, with one conductor per slot. Develop a progressive simplex lap winding diagram for the DC machine, mark the poles. Also show equiliser ring.*

Solution :

Analysing the Problem : This particular problem states us to develop a lap winding diagram for a DC machine having 28 armature conductors and 4 poles. Drawing sequence diagram helps us to analyse the flow of current and helps to locate the brushes.

Equalising Connection : It is a characteristic of lap winding that all conductors in any parallel path lies under one pair of pole. Each parallel path carries same current when it place exactly under the pole, but due to slight variation in air gap or magnetic property, there is inequalities in flux occurs. Hence, there is always slight imbalance of induced emf in the various parallel paths and also the brushes which carry unequal currents, some brushes will get over loaded. This results in poor commutation and affects the performance of the machine.

In order to overcome the problem of imbalance of induced emf, armature winding must be at equal potential and the difference in brush currents should be diminished. This requires that there should be a whole number of slots per pole air. For example, if there is slot under the centre of an N-pole which is connected to the conductor which are in the form of copper rings at armature back. These rings are called as equaliser ring. Hence, the function of equaliser ring is to avoid unequal distribution of the current at the brushes and also helps to get sparkles commutation.

Here,   m = number of rings

$$\text{Total number of tapping} = m \times \text{pairs of poles} = \frac{m \times P}{2}$$

Distance between adjacent tapping (T_g)

$$= \frac{\text{Total number of coils}}{\text{Total number of tapping}}$$

Given data :

Number of poles (P) = 4

Number of armature slots = number of conductors (Z) = 28

Step 1 : First calculate the pole pitch, front pitch Y_f and back pitch Y_b.

$$\text{Hence, pole pitch} = Z/P = 28/4 = 7$$

But　　　　　　pole pitch $= (Y_b + Y_f)/2$

Hence,　　　　$(Y_b + Y_f)/2 = 7$

$$(Y_b + Y_f) = 14 \qquad \text{... (i)}$$
$$(Y_b - Y_f) = 2 \qquad \text{... (ii)}$$

Solving Eq. (i) and (ii), we get

$$Y_b = 8 \text{ and } Y_f = 6$$

Here, both Y_b and Y_f are even.

Since, both Y_b and Y_f should be odd for lap winding in this particular problem Y_b and Y_f be considered as follows :

Hence,

$$Y_b = 7 \text{ and } Y_f = 5$$

Step 2 : Now, develop the winding table so as to establish the coil connection in various steps as shown in Table 3.3.

Table 3.3 : Winding Table

Sr. No.	At the Back $Y_b = 7$	At the Front $Y_f = 5$	Sr. No.	At the Back $Y_b = 7$	At the Front $Y_f = 5$
1	1 + 7 = 8	8 − 5 = 3	8	15 + 7 = 22	22 − 5 = 17
2	3 + 7 = 10	10 − 5 = 5	9	17 + 7 = 24	24 − 5 = 19
3	5 + 7 = 12	12 − 5 = 7	10	19 + 7 = 26	26 − 5 = 21
4	7 + 7 = 14	14 − 5 = 9	11	21 + 7 = 28	28 − 5 = 23
5	9 + 7 = 16	16 − 5 = 11	12	23 + 7 = 30(2)	30 − 5 = 25
6	11 + 7 = 18	18 − 5 = 13	13	25 + 7 = 32(4)	32 − 5 = 27
7	13 + 7 = 20	20 − 5 = 15	14	27 + 7 = 34(6)	34 − 5 = 29(1)

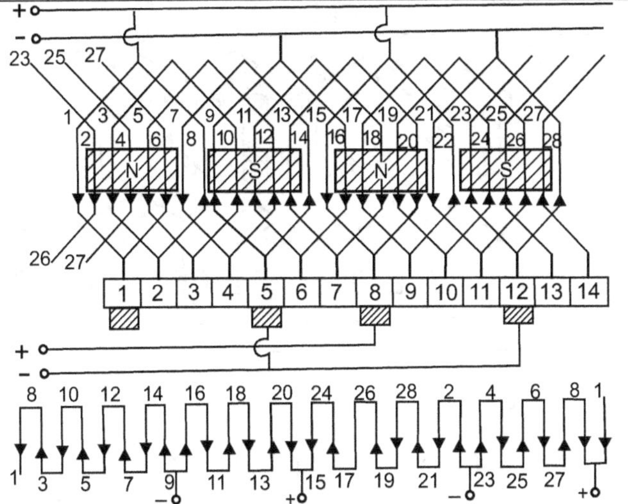

Fig. 3.15 : Winding diagram and Ring Diagram

Example 3.4 : A DC machine has 32 armature slots and 4 poles, with one conductor per slot. Develop a progressive duplex lap winding diagram for the DC machine.

Solution : Analysing the Problem : This particular problem states us to develop a duplex lap winding diagram for a DC machine having 32 armature conductors and 4 poles.

Step 1 : First calculate the pole pitch, front pitch Y_f and back pitch Y_b.

Hence,　　　　pole pitch $= Z/P = 32/4 = 8$

But　　　　　　pole pitch $= (Y_b + Y_f)/2$

Hence,　　　　$(Y_b + Y_f)/2 = 8$

Calculating Y_b as per the general procedure :

$$(Y_b + Y_f) = 16 \qquad \text{... (i)}$$
$$(Y_b - Y_f) = 2 \qquad \text{... (ii)}$$

Solving equations (i) and (ii), we get

$$Y_b = 9$$

But　　　　　　$Y_b = Y_f \pm 2m$

(where m = 1 for simplex, 2 for duplex and 3 for triplex)

$$(Y_b - Y_f) = 2 \times 2 = 4 \qquad \text{... (iii)}$$

Now, substituting the value of Yb in equation (iii)

Hence,　　　$Y_b = 9 \text{ and } Y_f = 5$

Step 2 : Now, develop the winding table so as to establish the coil connection in various slots as shown in Table 3.4.

Table 3.4 : Winding Table

At the Back $Y_b = 9$	At the Front $Y_f = 5$	At the Back $Y_b = 9$	At the Front $Y_f = 5$
1 + 9 = 10	10 − 5 = 5	3 + 9 = 12	12 − 5 = 7
5 + 9 = 14	14 − 5 = 9	7 + 9 = 16	16 − 5 = 11
9 + 9 = 18	18 − 5 = 13	11 + 9 = 20	20 − 5 = 15
13 + 9 = 22	22 − 5 = 17	15 + 9 = 24	24 − 5 = 19
17 + 9 = 26	26 − 5 = 21	19 + 9 = 28	28 − 5 = 23
21 + 9 = 30	30 − 5 = 25	23 + 9 = 32	32 − 5 = 27
25 + 9 = 34(2)	34 − 5 = 29	27 + 9 = 36(4)	36 − 5 = 31
29 + 9 = 38(6)	38 − 5 = 33(1)	31 + 9 = 40(8)	40 − 5 = 35(3)

Now, this winding table shows the connection of coils on both front and back pitches. Repeat the steps 1 to 20 as stated in solution of problem to draw the winding diagram. Fig. 3.16 shows the developed winding diagram for this particular problem.

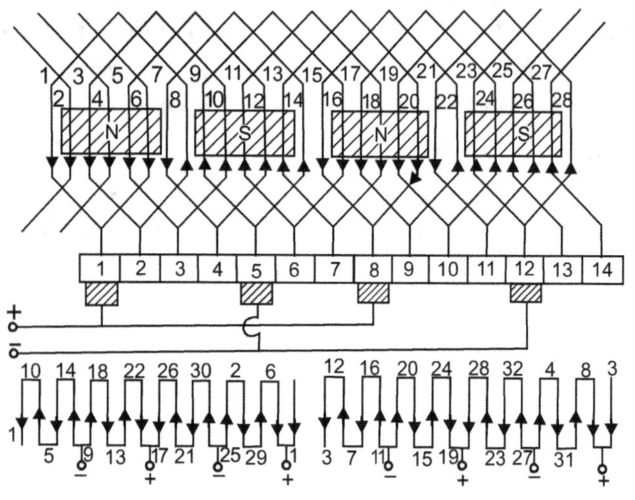

Fig. 3.16 : Winding diagram and Ring Diagram

Example 3.5 : Develop a double layer winding for a DC machine having 16 slots and 4 poles. Draw the sequence diagram and indicate the position of the brushes. Show the direction of induced emf and give equaliser connection.

Solution : Given data :

$$\text{Type of winding } = \text{ Progressive lap}$$
$$\text{Number of slots (s) } = 16$$
$$\text{Number of conductors (Z) } = 32$$
$$\text{Number of poles (P) } = 4$$
$$\text{Number of circuits (m) } = 1$$
$$\text{Number of layers } = 2$$

Calculation :

$$\text{Pole Pitch (Z/P) } = 8$$
$$Y_b = 9$$
$$Y_f = 7$$

Table 3.5 : Winding Table

At the Back	At the Front	At the Back	At the Front
$Y_b = 9$	$Y_f = 7$	$Y_b = 9$	$Y_f = 7$
1 + 9 = 10	10 − 7 = 3	17 + 9 = 26	26 − 7 = 19
3 + 9 = 12	12 − 7 = 5	19 + 9 = 28	28 − 7 = 21
5 + 9 = 14	14 − 7 = 7	21 + 9 = 30	30 − 7 = 23
7 + 9 = 16	16 − 7 = 9	23 + 9 = 32	32 − 7 = 25
9 + 9 = 18	18 − 7 = 11	25 + 9 = 34(2)	34 − 7 = 27
11 + 9 = 20	20 − 7 = 13	27 + 9 = 36(4)	36 − 7 = 29
13 + 9 = 22	22 − 7 = 15	29 + 9 = 38(6)	38 − 7 = 31
15 + 9 = 24	24 − 7 = 17	31 + 9 = 40(8)	40 − 7 = 33(1)

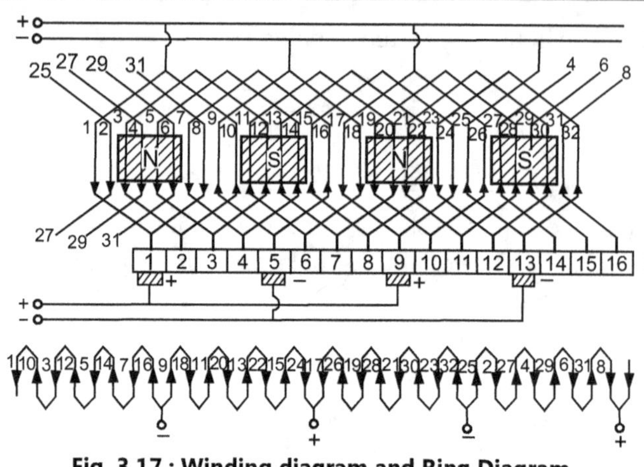

Fig. 3.17 : Winding diagram and Ring Diagram

Example 3.6 : Draw the armature winding of a DC machine with 2 poles, 18 slots and single layer retrogressive wave. Draw the sequence diagram and show the position of brush and direction of induced emf.

Solution : Given data :

$$\text{Type of winding } = \text{ Retrogressive wave}$$
$$\text{Number of slots } = 18$$
$$\text{Number of conductor (Z) } = 18$$
$$\text{Number of poles (P) } = 2$$
$$\text{Number of circuits (m) } = 1$$
$$\text{Number of layers } = 1$$
$$\text{Calculation : } \quad \text{Pole pitch (Z/P) = 9}$$
$$Y_b = 7$$
$$Y_f = 9$$

Now, develop the winding table so as to establish the coil connection in various slots as shown in Table 3.6

Table 3.6 : Winding Table

At the Back	At the Front	At the Back	At the Front
$Y_b = 7$	$Y_f = 9$	$Y_b = 7$	$Y_f = 9$
1 + 7 = 8	8 + 9 = 17	9 + 7 = 16	16 + 9 = 25(7)
17 + 7 = 24(6)	6 + 9 = 15	7 + 7 = 14	14 + 9 = 23(5)
15 + 7 = 22(4)	4 + 9 = 13	5 + 7 = 12	12 + 9 = 21(3)
13 + 7 = 20(2)	2 + 9 = 11	3 + 7 = 10	10 + 9 = 19(1)
11 + 7 = 18	18 + 9 = 27(9)		

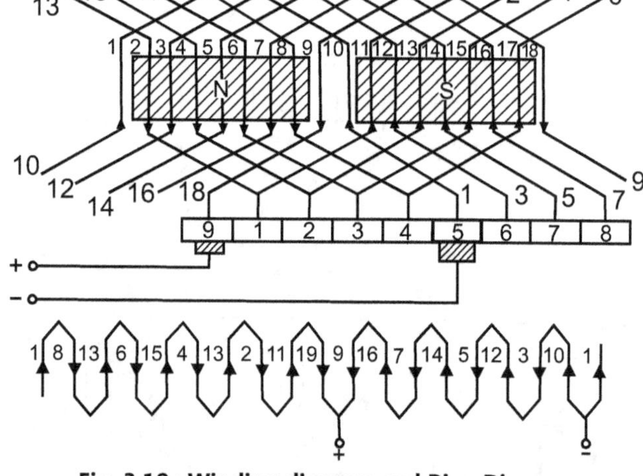

Fig. 3.18 : Winding diagram and Ring Diagram

Example 3.7 : *Draw the armature winding of a DC machine with 4 poles, 18 slots and double layer simplex retrogressive wave. Show the position of brush and direction of induced emf.*

Solution : Given data :

$$\text{Type of winding} = \text{Retrogressive wave}$$
$$\text{Number of slots} = 18$$
$$\text{Number pf conductor (Z)} = 36$$
$$\text{Number poles (P)} = 4$$
$$\text{Number of circuits (m)} = 1$$
$$\text{Number of layers} = 2$$

Calculation :

$$\text{Pole pitch (Z/P)} = 9$$
$$Y_b = 9$$
$$Y_f = 9$$

Note : Here, in this problem the commutator pitch is fraction which results a coil to be made electrical dummy, i.e., a coil will be connected at the back end and left open in the front end connection.

Therefore, the number of conductors to be considered is 34(Z = 34).

Now, develop the winding table so as to establish the coil connection in various slots as shown in Table 3.7

Table 3.7 : Winding Table

At the Back $Y_b = 9$	At the Front $Y_f = 9$	At the Back $Y_b = 9$	At the Front $Y_f = 9$
1 + 9 = 10	10 + 9 = 19	27 + 9 = 36(2)	2 + 9 = 11
19 + 9 = 28	28 + 9 = 37(3)	11 + 9 = 20	20 + 9 = 29
3 + 9 = 12	12 + 9 = 21	29 + 9 = 38(4)	4 + 9 = 13
21 + 9 = 30	30 + 9 = 39(5)	13 + 9 = 22	22 + 9 = 31
5 + 9 = 14	14 + 9 = 23	31 + 9 = 40(6)	6 + 9 = 15
23 + 9 = 32	32 + 9 = 41(2)	15 + 9 = 24	24 + 9 = 33
7 + 9 = 16	16 + 9 = 25	33 + 9 = 42(8)	8 + 9 = 17
25 + 9 = 34	34 + 9 = 43(9)	17 + 9 = 26	26 + 9 = 35(1)
9 + 9 = 18	18 + 9 = 27		

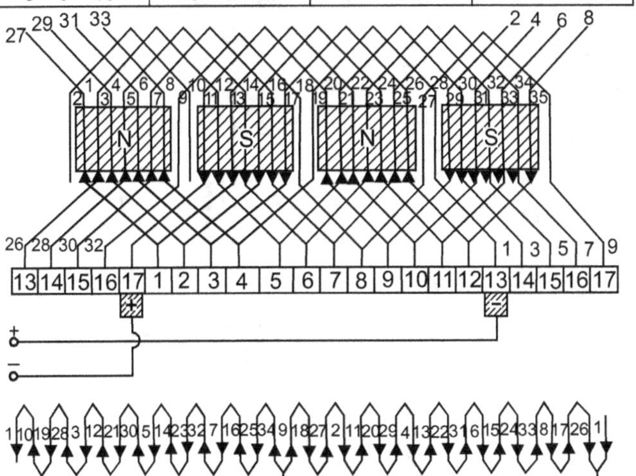

Fig. 3.19 : Winding diagram and Ring Diagram

Example 3.8 : *Develop a single layer simplex progressive wave winding for a dc machine with 4 poles and 24 slots. Show the position of brush and direction of induced emf. Draw the sequence diagram.*

Analysing the Problem : *This particular problem states us to develop a single layer progressive wave winding diagram for a DC machine having 24 armature conductors and 4 poles. Drawing sequence diagram helps us to analyse the flow of current and helps to locate the brushes.*

Solution : Given data:

$$\text{Number of poles (P)} = 4$$

In single layer, number of conductors (Z) = slots = 24

Step 1 : First calculate the pole pitch, front pitch Y_f and back pitch Y_b.

Hence,   pole pitch (Z/P) = 24/4 = 6

Therefore,

$$Y_b = 7 \text{ and } Y_f = 7$$

Step 2 : Now, develop the winding table so as to establish the coil connection in various slots as shown in Table 3.8

Table 3.8 : Winding Table

At the Back $Y_b = 7$	At the Front $Y_f = 7$	At the Back $Y_b = 7$	At the Front $Y_f = 7$
1 + 9 = 8	8 + 7 = 15	13 + 7 = 20	20 + 7 = 27(3)
15 + 7 = 22	22 + 7 = 29	3 + 7 = 10	10 + 7 = 17
5 + 7 = 12	12 + 7 = 19	17 + 7 = 24	24 + 7 = 31(3)
19 + 7 = 26(2)	2 + 7 = 9	7 + 7 = 14	14 + 9 = 21
9 + 7 = 16	16 + 7 = 23	21 + 7 = 28(4)	4 + 7 = 11
23 + 7 = 30(6)	6 + 7 = 13	11 + 7 = 18	18 + 7 = 25

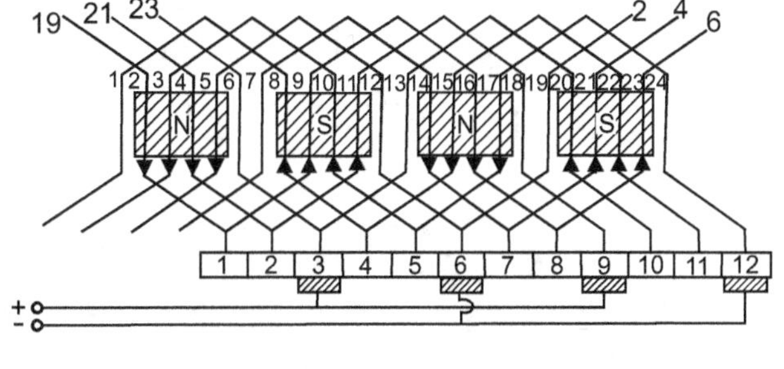
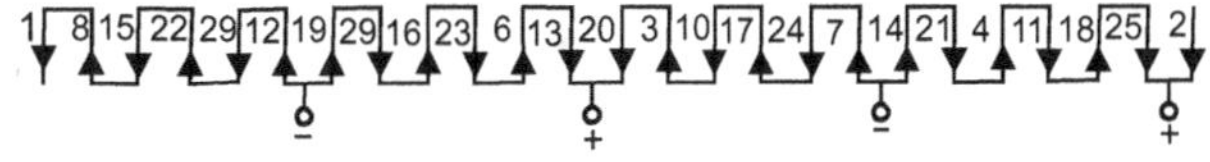

Fig. 3.20 : Winding diagram and Ring Diagram

Example 3.9 : *Draw the winding diagram of a D C Machine with 4 poles, 14 slots, progressive, double layer lap winding. Show the position of brushes and direction of induced emf.*

Solution : Number of poles = 4 ; Number of slots = 14, Number of conductors = 14 x 2 = 28

Pole pitch = Number of conductors/pole = 28/4 = 7

We have pole pitch = $(Y_b + Y_f) / 2 = Y_p$

Hence

$$(Y_b + Y_f) = 14$$
$$(Y_b - Y_f) = 2$$

Solving above equations $Y_b = 8$ and $Y_f = 6$

$$\text{back pitch } y_b = 2c/p \pm k$$

For lap winding both Y_b and Y_f must be odd and differ by 2 Satisfying the above condition $Y_b = 7$ and $Y_f = 5$ (Winding diagram and ring diagrams are shown below)

Table 3.9 : Winding Table

At the Back $Y_b = 7$	At the Front $Y_f = 5$	At the Back $Y_b = 7$	At the Front $Y_f = 5$
Coil Side → connected to Coil Side	Coil Side → Connected to Coil Side	Coil Side → Connected to Coil Side	Coil Side → Connected to Coil Side
1 + 7 = 8	8 – 5 = 3	17 + 7 = 24	24 – 5 = 19
3 + 7 = 10	10 – 5 = 5	19 + 7 = 26	26 – 5 = 21
5 + 7 = 12	12 – 5 = 7	21 + 7 = 28	28 – 5 = 23
7 + 7 = 14	14 – 5 = 9	33 + 7 = 30 (2)	30 – 5 = 25
9 + 7 = 16	16 – 5 = 11	25 + 7 = 32(4)	32 – 5 = 27
11 + 7 = 18	18 – 5 = 13	27 + 7 = 34(6)	34 – 5 = 29(1)
13 + 7 = 20	20 – 5 = 15		
15 + 7 = 22	22 – 5 = 17		

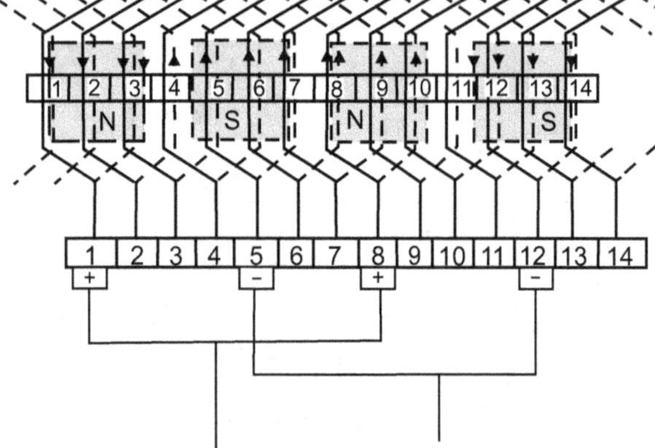

Fig. 3.21 : Winding diagram

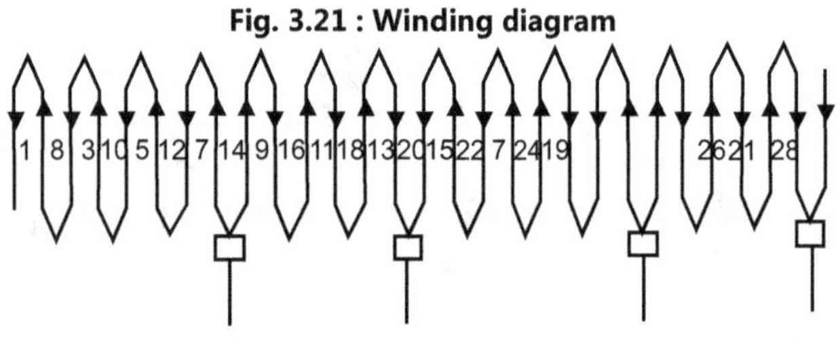

Fig. 3.22 : Ring Diagram

Example 3.10 : *Develop the single layer winding for a D C machine having 32 armature conductors and 4 poles. Mark the poles Draw the sequence diagram, indicate the position of the brushes and the direction of induced emf and show the equiliser connections.*

Solution : Number of conductors = 32 Pole pitch = 32/4 = 8;

$$\text{pole pitch} = (Y_b + Y_f) / 2 = Y_p$$

Hence

$$(Y_b + Y_f) = 16 \text{ and } (Y_b - Y_f)$$
$$= 2 \text{ hence } Y_b = 9 \text{ and } Y_f = 7$$

(Winding diagram and ring diagrams are shown below)

Table 3.10 : Winding Table

At the Back $Y_b = 9$	At the Front $Y_f = 7$	At the Back $Y_b = 9$	At the Front $Y_f = 7$
Coil Side → Connected to Coil Side	Coil Side → Connected to Coil Side	Coil Side → Connected to Coil Side	Coil Side → Connected to Coil Side
1 + 9 = 10	10 – 7 = 3	17 + 9 = 26	26 – 7 = 19
3 + 9 = 12	12 – 7 = 5	19 + 9 = 28	28 – 7 = 21
5 + 9 = 14	14 – 7 = 7	21 + 9 = 30	30 – 7 = 23
7 + 9 = 16	16 – 7 = 9	23 + 9 = 32	32 – 7 = 25
9 + 9 = 18	18 – 7 = 11	25 + 9 = 34(2)	34 – 7 = 27
11 + 9 = 20	20 – 7 = 13	27 + 9 = 36(4)	36 – 7 = 29
13 + 9 = 22	22 – 7 = 15	29 + 9 = 38(6)	38 – 7 = 31
15 + 9 = 24	24 – 7 = 17	31 + 9 = 40(8)	40 – 7 = 33(1)

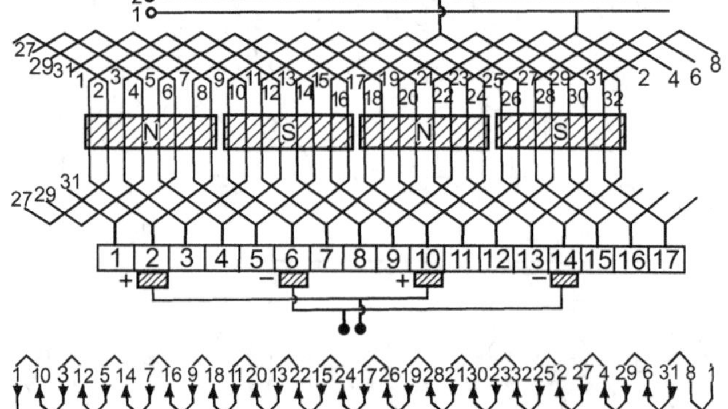

Fig. 3.23 : Winding diagram and Ring Diagram

AC WINDINGS

3.14 INTRODUCTION

- Armature winding of alternators is different form that of d.c machines. Basically three phase alternators carry three sets of windings arranged in the slots in such as way that there exists a phase difference of 120 deg between the induced emf's in them.
- In a d.c machine, winding is closed while in alternators winding is open i.e., two ends of each set of winding is brought out.
- In three phase alternators, the six terminals are brought out which are finally connected in star or delta and then the three terminals are brought out.
- Each set of windings represent winding per phase and induced e.m.f, each set is called induced e.m.f per phase denoted as Eph.
- All the coils used for one phase must be connected in such a way that their e.m.f.s help each other.

3.14.1 Armature Windings for AC Machines

1. Similarities between DC and AC Machines

Sr. No.	Parameter	DC – Machine	AC – Machine (Sync)
1.	DC magnetic-flux	Exists	Exists
2.	Single and Double Layers	Exists	Exists
3.	Lap and Wave Wound	Exists	Exists
4.	Progressive and Retrogressive	Exists	Exists
5.	Full Pitch and Short pitch	Exists	Exists
6.	Wave form of Air-gap flux	Non-Sinusoidal	Non-Sinusoidal
7.	Voltage induced in armature conductors	AC (DC communication)	AC

2. Differences between DC and AC Machines

Sr. No.	Parameter	DC – Machine	AC – Machine (Sync)
1.	Armature	Rotating only	Mostly Stationary (Except in very small ratings)
2.	Terminal Voltage	DC	AC
3.	Wave form of Voltage	Constant DC	Nearly Sinusoidal
4.	Harmonics	Do not Exist	Exists
5.	Effect of Chording	Changes the voltage	Changes magnitude and harmonics

3.15 WINDING TERMINOLOGY

- **Conductor:** The part of the wire, which is under the influence of the magnetic field and responsible for the induced e.m.f is called active length of the conductor. The conductor are placed in the armature slots.
- **Turn:** A conductor in one slot, when connected to a conductor in another slot forms a turn. So two conductors constitute a turn. This is shown in Fig. 3.24 below.

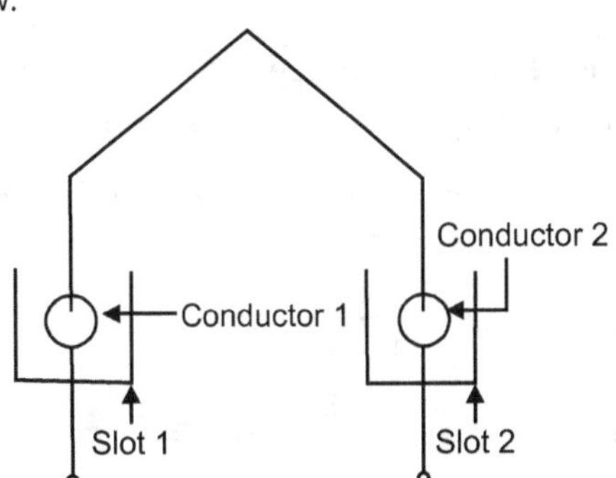

Fig. 3.24 : Conductor and coil diagram

- **Coil:** As there are number of turns, for simplicity the number of turns are grouped together to form a coil Such a coil is called multi-turn coil. A coil may consist of single turn called single turn coil. The Fig. 3.25 shows a multi-turn coil.
- **Coil Side:** Coil consists of many turns. Part of the coil in each slot is called coil side as shown in the Fig. 3.25 below

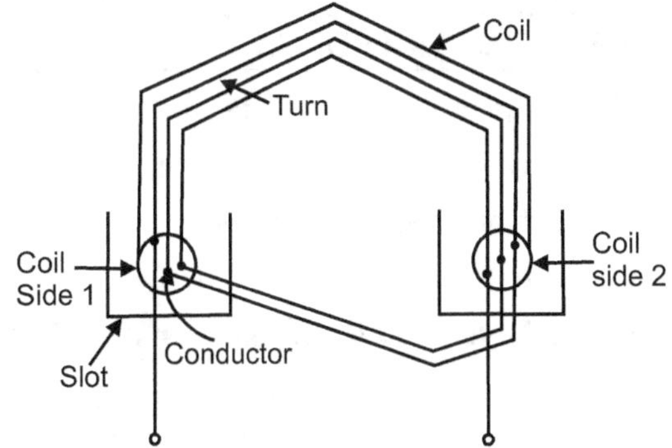

Fig. 3.25 : Coil and coil sides diagram

- **Pole Pitch :** It is centre to centre distance between the two adjacent poles. We have seen that for one rotation of the conductors, 2 poles are responsible for 360° electrical of e.m.f., 4 poles are responsible for 720° electrical of e.m.f and so on. So 1 pole is responsible for 180° electrical of induced e.m.f
- **Key Point:** so 180° electrical is also called one pole pitch. Practically how many slots are under one pole which are responsible for 180° electrical, are measured

to specify the pole pitch. e.g. Consider 2 pole, 18 slots armature of an alternator. Then under 1 pole there are 18/2 i.e. 9 slots. So pole pitch is 9 slots or 180° electrical. This means 9 slots are responsible to produce a phase difference of 180° between the e.m.f.s induced in different conductors. This number of slots/pole is denoted a 'n'.

Pole pitch = 180° electrical = slots per pole (no. of slots/P) = n

- **Slot Angle (β) :** The phase difference contributed by one slot in degrees electrical is called slot angle β.

As slots per pole contributes 180° electrical which is denoted as 'n', we can write,

$$1 \text{ slot angle} = \frac{180 \text{ deg}}{n}$$

In the above example, $n = \dfrac{18}{2}$ while $\beta = \dfrac{180 \text{ deg}}{n} = 20$ deg

Note : This means that if we consider an induced e.m.f in the conductors which are placed in the slots which are adjacent to each other, there will exist a phase difference of β° in between them.

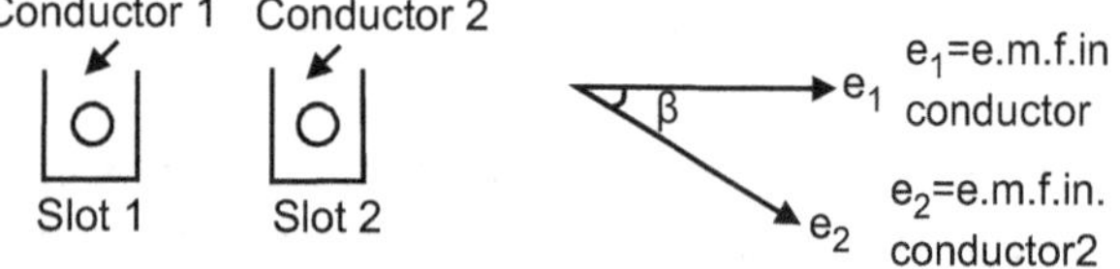

Fig. 3.26 : Adjacent slots and indication of phase difference

3.16 CLASSIFICATION OF AC WINDING

In general armature winding is calssified as,

1. Single layer and double layer winding
2. Full pitch and short pitch winding
3. Concentrated and distributed winding.

3.16.1 Single Layer and Double Layer Winding

- If a slot consists of only one coil side, winding is said to be single layer. This is shown in the Fig. 3.27.

- While there are two coil sides per slot, one at the bottom and one at the top the winding is called double layer as shown in the Fig. 3.27.

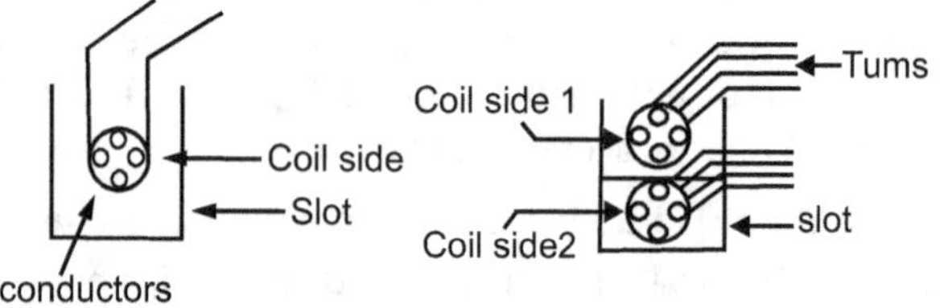

Fig. 3.27 : Single layer and double layer winding

3.16.2 Full Pitch Winding

- As seen earlier, one pole pitch is 180° electrical. The value of 'n', slots per pole indicates how many slots are contributing 180° electrical phase difference. So if coil side in one slot is connected to a coil side in another slot which is one pole pitch distance away from first slot, the winding is said to be **full pitch winding** and coil is called **full pitch coil.**

For Example : In 2 pole, 18 slots alternator, the pole pitch is n = 18/2 = 9 slots. So if coil side in slot No.1 is connected to coil side in slot No. 10 such that two slots No.1 and No. 10 are one pole pitch or n slots or 180° electrical apart, the coil is called full pitch coil.

- **Coil Span :** It is the distance on the periphery of the armature between two coil sides of a coil. It is usually expressed in terms of number of slots or degrees electrical. So if coil span is 'n' slots or 180° electrical the coil is called full pitch coil. This is shown in the Fig. 3.28.

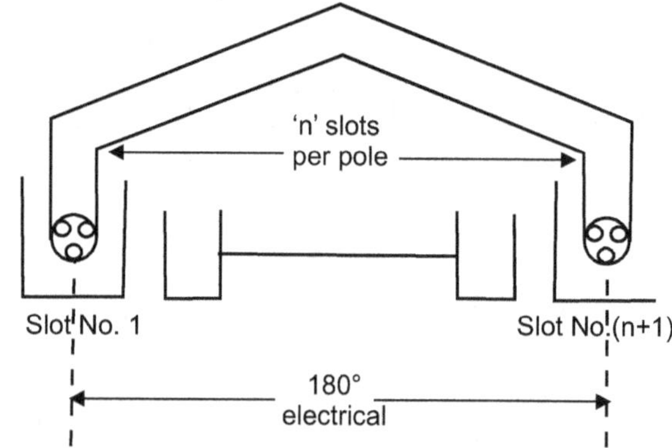

Fig. 3.28 : Full pitch winding

3.16.3 Short Pitch Winding

- As against this if coils are used in such a way that coil span is slightly less than a pole pitch i.e., less than 180° electrical, the coils are called, short pitched coils or fractional pitched coils.

- Generally coils are shorted by one or two slots.

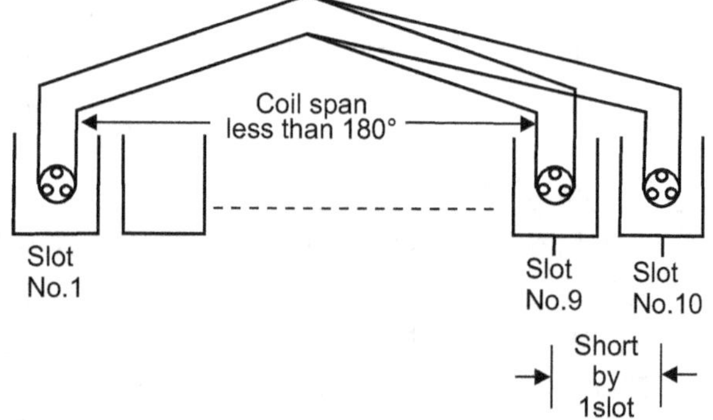

Fig. 3.29 : Short pitch winding

- So in 18 slots, 2 pole alternator instead of connecting a coil side in slot No.1 to slot No. 10, it is connected to a coil side in slot No.9 or slot No. 8, coil is said to be short pitched coil and winding is called short pitch winding.

3.17 TYPES OF DOUBLE LAYER WINDING

3.17.1 Integral Slot Winding

In an integral slot winding the number of slots/pole/phase 'm', and slots/pole are integer.

Example : S = 24, P = 4; 3 phase.

Since, $m = \dfrac{24}{4 + 3} = 2$ and $s/p = \dfrac{24}{4} = 6$ are integers, the winding is of integral slot type.

An integral slot winding can be of full of short pitch type.

3.17.2 Fractional Slot Winding

In an integral slot winding the number of slots/pole/phase 'm', and slots/pole are integer. i.e., If Slots/Pole/Phase (SPP) is not a whole number (Example: for a 66 slots, 4 pole, 3ph armature, S/P/P = 66/4/3 = 5.5) An fractional slot winding is always short pitch windings.

3.18 PHASE SPREAD

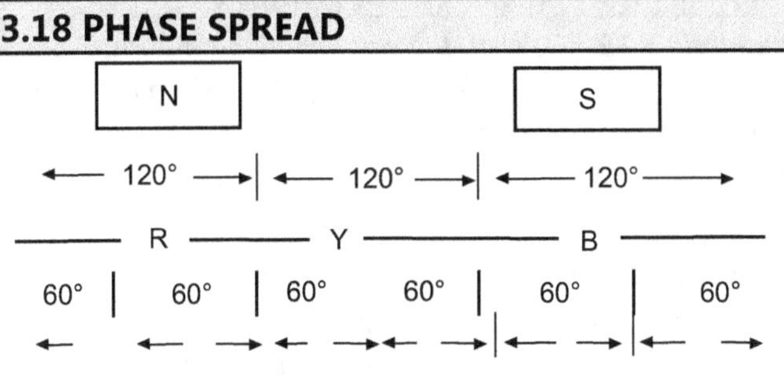

Fig. 3.30 : Phase spread

120° Phase spread or wide spread winding.

60° phase spread or narrow spread winding

- In a 120 deg phase spread winding, each phase coils, pole occupies a space equivalent to 120° electrical as shown in the Fig. 3.30.

- Thus the space sequence of the phase band or phase winding arrangement sequence is RYB for a phase sequence of RBY.

- In a 60° phase spread winding, each phase coils occupies a space equivalent to 60° under each pole as shown in the Fig. 3.30.

- Thus the space sequence of the phase band or phase winding arrangement sequence is RBY for a phase sequence of RYB.

- Between 120° and 60° phase winding, 60° phase spread winding is much used.

- As the distribution factor of 60° phase spread is greater than 120° phase spread.

- The minimum value of distribution factor for fundamental is 0.955 for 60° phase spread and 0.827 for 120° phase spread

3.19 STARTING OF PHASES

- To calculate the starting of the phases, consider the phase sequence as reference.

- Angle between R and Y is 120°, R and B is 240°

- Let the starting of the R_{phase} be the I slot

3.20 ELIMINATION OF N^{TH} HARMONICS

- The short pitch winding is more popular compared to full pitch winding as it saves copper at overhang and helps to reduce or eliminate the harmonics.

- A harmonic is said to be eliminated, i.e., the emf induced due to the harmonics $E_n = 4.44$ is zero.

$$K_p = 1 \text{ (full pitch)}$$
$$K_p \quad 1 \text{ (Short pitch)}$$

This happens when the pitch factor of Nth Harmonic K_p represents zero.

$$K_{pn} = \cos \frac{n\alpha}{2}$$

$$\frac{n\alpha}{2} = \cos^{-1}(0) = 90°$$

$$n\alpha = 180°$$

$$\alpha = \frac{180°}{n} \quad \text{where } n = \text{No. of harmonics}$$

Example 3.11 : Draw the developed single layer lap winding diagram of a 3 phase star connected AC machine with 4 poles and 24 conductors. The winding is to be short pitched by one slot.

Solution :

Analysing the Problem : This particular problem states us to develop a single layer lap winding diagram for an AC machine having 24 conductors and 4 poles.

Solution : Given data :

Number of poles = 4

Number of slots = S = Z = number pf conductors = 24

Step 1 : First calculate all the necessary details like pole pitch, slots/pole, slots/pole/phase, coil span, etc. as given below :

- Pole pitch = number of slots/pole = (24/4) = 6

- Number of slots/pole/phase = (6/3) = 2

- Number of coils/pole/phase = 12/12 = 1 (hence winding is balanced)

- Slot angle = (180/pole pitch) = (180/6) = 30°
- Winding pitch = 180 – (number of slots shorted × slot angle) = 180 – (1 × 30) = 150
- Coil span = (winding pitch/slot angle) = (150/30) = 5
- The winding is balanced and one coil is fitted in each pole phase. Hence, 1st conductor is connected to 6th conductor (1 + 5 = 6).
- The phase R starts from slot number 1, i.e., R_S = 1.
- The phase Y starts after 120°, i.e., Y_S = 1 + (120slot angle) = 1 + (120/30) = 5.
- The phase B starts after 240°, i.e., B_S = 1 + (240/slot angle) = 1 + (240/30) = 9.

Step 2 : Now, develop the winding table so as to establish the coil connection in various slots as shown in Table 3.11

Table 3.11 : Winding Table

	1st Pole	2nd Pole	3rd Pole	4th Pole
R phase	1 + 5 = 6	7 + 5 = 12	13 + 5 = 18	19 + 5 = 24
B phase	3 + 5 = 8	9 + 5 = 14	15 + 5 = 20	21 + 5 = 26(2)
Y phase	5 + 5 = 10	11 + 5 = 16	17 + 5 = 22	23 + 5 = 28(4)

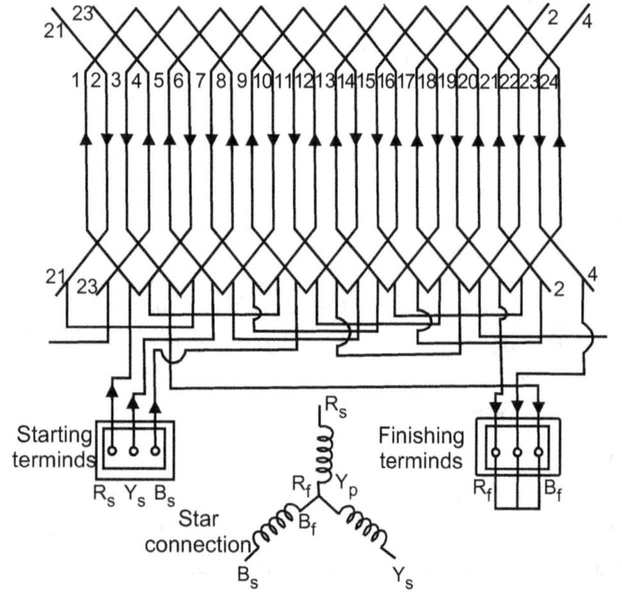

Fig. 3.31 : Winding diagram

Example 3.12 : *Draw the developed winding diagram for the following details : 3 phase AC double layer lap winding, slots = 12, poles = 4 and phase sequence = R Y B.*

Solution : Pole pitch = S/P = 3

Slots per pole per phase = pole pitch/phase = 1

Slot angle = 180°/pole pitch = 60°

$R_S = 1^{TCS}$ (TCS = Top Coil Side in particular slot)

$Y_S = 1 + (120°/60°) = 3^{TCS}$

$B_S = 1 + (240°/60°) = 5^{TCS}$

Table 3.12 : Winding Table

Poles/Phases	R Phase	B Phase	Y Phase
P_1	(R_S)1 + 3 = 4'	(B_F)2 + 3 = 5'	(Y_S)3 + 3 = 6'
P_2	4 + 3 = 7'	(B_S)5 + 3 = 8'	6 + 3 = 9'
P_3	7 + 3 = 10'	8 + 3 = 11'	9 + 3 = 12'
P_4	(R_F)10 + 3 = 13'	11 + 3 = 14'	(Y_F)12 + 3 = 15'

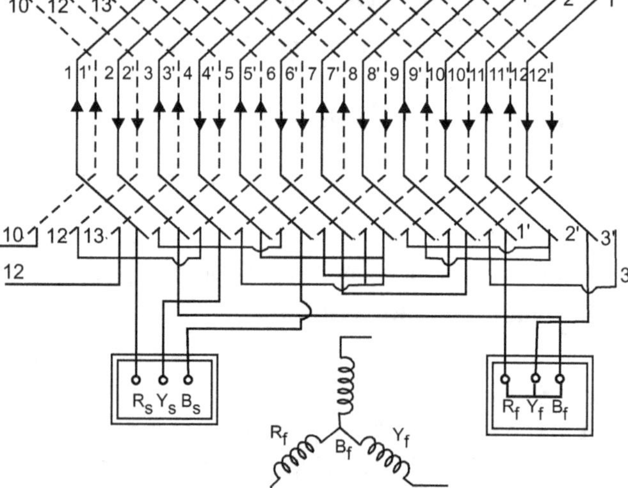

Fig. 3.32 : Winding diagram

Example 3.13 : *Develop the winding diagram for a 3 phase AC machine with 18 slots and 6 poles. Assume it is star connected and phase sequence is RYB.*

Solution :

Pole pitch = S/P = 18/6 = 3

Slots per pole per phase = pole pitch/phase = 1

Slot angle = 180°/pole pitch = 60°

$R_S = 1^{TCS}$ (TCS = Top Coil Side in particular slot)

$Y_S = 1 + (120°/60°) = 3^{TCS}$

$B_S = 1 + (240°/60°) = 5^{TCS}$

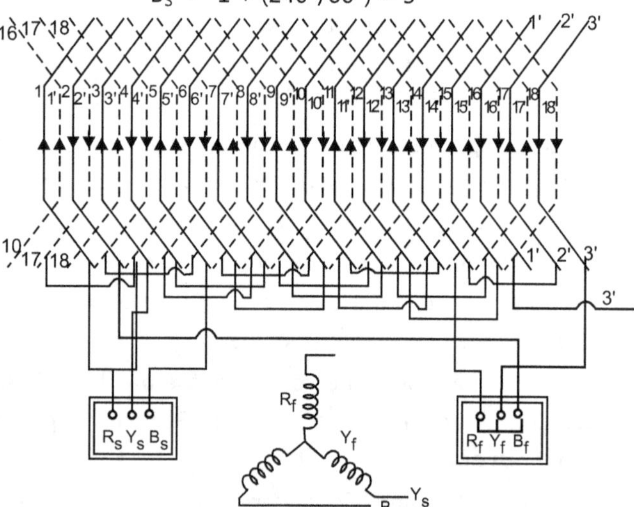

Fig. 3.33 : Winding diagram

Example 3.14 : *Design and draw the developed 3 phase AC lap winding diagram for 24 slots, double layer lap connected, 4 poles and one slot to be chorded. Also show the winding in star connection.*

Solution :

Pole pitch $= S/P = 6$

Slots per pole per phase = pole pitch/phase = 2

Slot angle = 180°/pole pitch = 30°

$R_S = 1^{TCS}$ (TCS = Top Coil Side in particular slot)

$Y_S = 1 + (120°/30°) = 5^{TCS}$

$B_S = 1 + (240°/30°) = 9^{TCS}$

Actual coil span = pole pitch – number of coils to be chorded = 6 – 1 = 5

Now, develop the winding table so as to establish the coil connection in various slots as shown in Table 3.12

Table 3.13 : Winding Table

Poles/ Phases	R Phase	B Phase	Y Phase
P1	$(R_S)1 + 5 = 6'$ $2 + 5 = 7'$	$3 + 5 = 8'$ $4 + 5 = 9'$	$5 + 5 = 10'$ $6 + 5 = 11'$
P2	$7 + 5 = 12'$ $8 + 5 = 13'$	$9 + 5 = 14'$ $10 + 5 = 15'$	$11 + 5 = 16'$ $12 + 5 = 17'$
P3	$13 + 5 = 18'$ $14 + 5 = 19'$	$15 + 5 = 20'$ $16 + 5 = 21'$	$17 + 5 = 22'$ $18 + 5 = 23'$
P4	$19 + 5 = 24'$ $20 + 5 = 25'(1)'$	$21 + 5 = 26'\ (2')$ $22 + 5 = 27'(3)'$	$23 + 5 = 28'(4')$ $24 + 5 = 29'(5')$

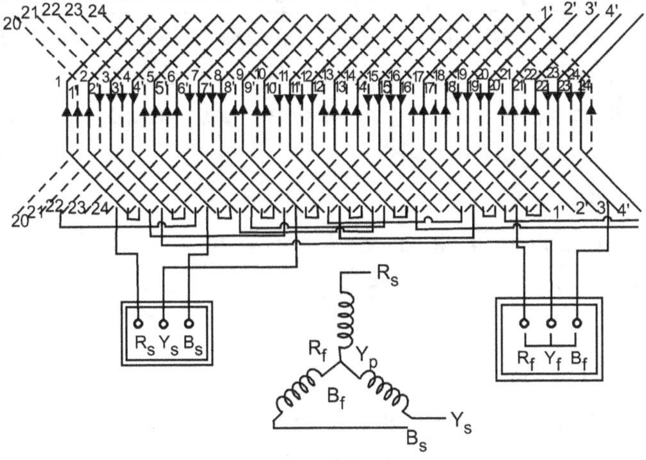

Fig. 3.34 : Winding diagram

Example 3.15 : *Develop and draw the wave winding diagram for the following details: Slots = 12, poles = 4, single layer, phase sequence is RYB and delta connected.*

Solution : Pole pitch = S/P = 3

Slots per pole per phase = pole pitch / phase = 1

Slot angle = 180°/pole pitch = 60°

$R_S = 1^{TCS}$

$Y_S = 1 + (120°/60°) = 3^{TCS}$

$B_S = 1 + (240°/60°) = 5^{TCS}$

The coil span is 6, hence the Y_B and Y_F = 6.

T_S conductor in 1^{TCS} = 1 (TCS = Top Coil Side in particular slot)

Y_S conductor in 3^{TCS} = 5

B_S conductor in 5^{TCS} = 9

Table 3.14 : Winding Table

R_S 1 + 6 = 7	7 + 6 = 13	13 + 6 = 19	19 + 6 = 25(1)
2 + 6 = 8	8 + 6 = 14	14 + 6 = 20	R_F 20 + 6 = 26(2)
Y_S 5 + 6 = 11	11 + 6 = 17	17 + 6 = 23	23 + 6 = 29(5)
6 + 6 = 12	12 + 6 = 18	18 + 6 = 24	Y_F 24 + 6 = 30(6)
B_S 9 + 6 = 15	15 + 6 = 21	21 + 6 = 27(3)	3 + 6 = 9
10 + 6 = 16	16 + 6 = 22	22 + 6 = 28(4)	B_F 4 + 6 = 10

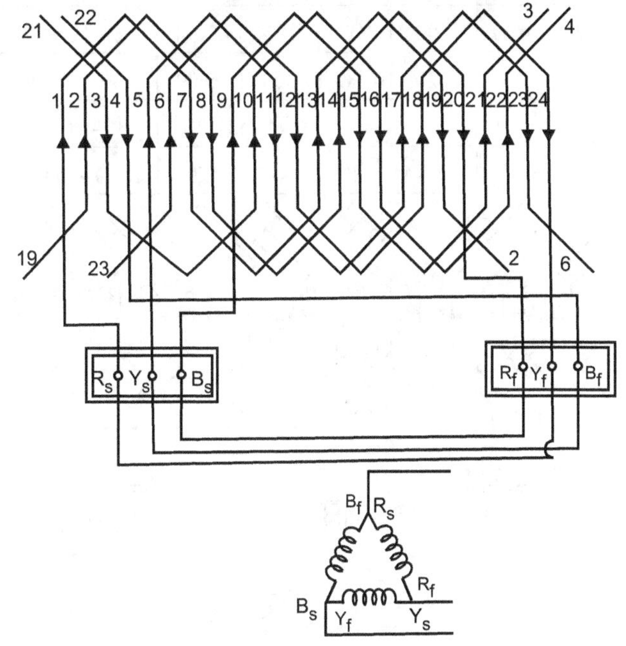

Fig. 3.35 : Winding diagram

Example 3.16 : *Develop the wave winding for the stator of an induction motor having 24 slots, 4 poles, RYB phase sequence and double layer.*

Solution :

Pole pitch $= S/P = 4$

Slots per pole per phase = pole pitch/phase = 2

Slot angle = 180°/pole pitch = 30°

$R_S = 1^{TCS}$ (TCS = Top Coil Side in particular slot)

$Y_S = 1 + (120°/30°) = 5^{TCS}$

$B_S = 1 + (240°/30°) = 9^{TCS}$

The total winding pitch $Y = Y_b + Y_F$

$\qquad = 12 \times 2 = 24$ coil

$\qquad Y_B = 13$ and $T_F = 11$

T_S conductor in $1^{TCS} = 1$ (TCS = Top Coil Side in particular Slot)

Y_S conductor in $5^{TCS} = 9$

B_S conductor in $9^{TCS} = 17$

Table 3.15 : Winding Table

R_S 1 + 13 = 14	14 + 11 = 25	25 + 13 = 28	38 + 11 = 49(1)
3 + 13 = 16	16 + 11 = 27	27 + 13 = 40	40 + 11 = 51(3)
4 − 13 = 39	39 − 11 = 28	28 − 13 = 15	15 − 11 = 4
2 − 13 = 37	37 − 11 = 26	26 − 13 = 13	R_F 13 − 11 = 2
Y_S 9 + 13 = 22	22 + 11 = 33	33 + 13 = 46	46 + 11 = 57(9)
11 + 13 = 24	24 + 11 = 35	35 + 13 = 48	48 + 11 = 59(11)
12 − 13 = 47	47 − 11 = 36	36 − 13 = 23	23 − 11 = 12
10 − 13 = 45	45 − 11 = 34	34 − 13 = 21	Y_F 21 − 11 = 10
B_S 17 + 13 = 30	30 + 11 = 41	41 + 13 = 54(6)	6 + 11 = 17
19 + 13 = 32	32 + 11 = 43	43 + 13 = 56(8)	8 + 11 = 19
20 − 13 = 7	7 − 11 = 44	44 − 13 = 31	31 − 11 = 20
18 − 13 = 5	5 − 11 = 42	42 − 13 = 29	B_F 29 − 11 = 18

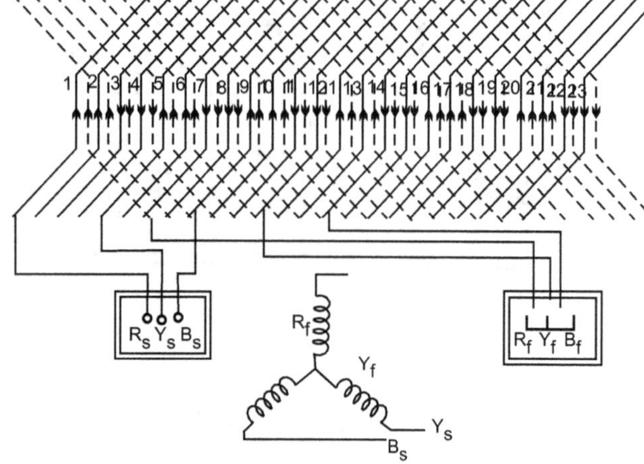

Fig. 3.36 : Winding diagram

Example 3.17 : *Design and draw the developed winding diagram of an alternator with following details: No of poles = 2 no. of phases = 3, No. of slots = 24, single layer lap winding, short pitched by one slot.*

Solution :

No. of poles = 2; No. of conductors = 24;

Pole pitch = 24/2 = 12; no of slots/pole /phase =24/ (2x3) = 4

No. of coils = 24/2 = 12

No of coils/pole/phase = 12/(2 × 3) = 2

Slot angle = 180/pole pitch = 180/12 = 15°

Winding pitch = 180 − (slot angle × no of slots shorted)

= 180 − 1 × 15 = 165

Hence coil span = 165° = 11 slots

Connections: R_S = 1, Y_S = 1 + 120/15 = 9;

B_S = 1 + 240/15 = 17

Table 3.16 : Winding Table

Phase	1st Pole	2nd Pole
R	1 + 11 = 12	13 + 11 = 24
	3 + 11 = 14	15 + 11 = 26 (2)
B	5 + 11 = 16	17 + 11 = 4
	7 + 11 = 18	19 + 11 = 6
Y	9 + 11 = 20	21 + 11 = 8
	11 + 11 = 22	23 + 11 = 10

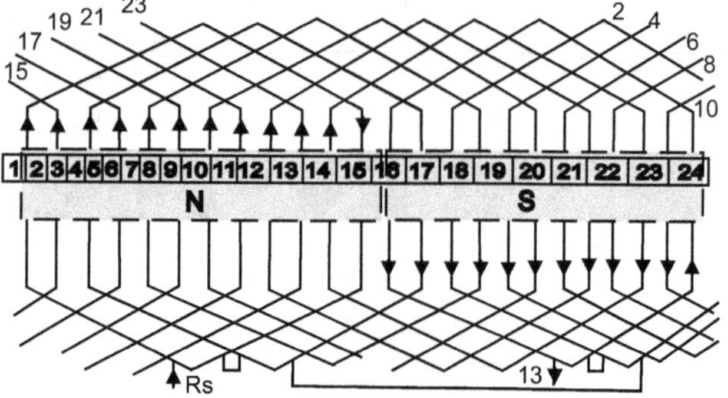

Fig. 3.37 : Winding diagram

Example 3.18 : *Design and draw the developed winding diagram of an alternator with following details: No of poles = 4 no. of phases = 3, No. of slots = 24, single layer wave winding, delta connected.*

Solution :

No. of poles = 4; No. of conductors = 24;

Pole pitch = 24/4 = 6; no of slots/pole /phase =24/ (4 × 3) = 2

No. of coils = 24/2 = 12

Slot angle = 180/pole pitch = 180/6 = 30°

Winding pitch = 180 − (slot angle)

$\qquad\qquad$ = 180 − 30= 150

Hence coil span = 180° / 30 = 6 slots

$\qquad Y_b = 6$ and $Y_f = 6$

Connections: $R_S = 1$, $Y_S = 1 + 120/30 = 5$;

$\qquad B_S = 1 + 240/30 = 9$

Table 3.17 : Winding Table

Phase R	1 + 6 = 7	7 + 6 = 13
	13 + 6 = 19	19 + 6 = 25 (1)
	(1 + 1) + 6 = 8	8 + 6 = 17
	14 + 6 = 20	
B	9 + 6 = 15	15 + 6 = 21
	21 + 6 = 27(3)	3 + 6 = 9
	10 + 6 = 16	16 + 6 = 22
	22 + 6 = 28 (4)	
Y	5 + 6 = 11	11 + 6 = 17
	17 + 6 = 23	23 + 6 = 29 (5)
	6 + 6 = 12	12 + 6 = 18
	18 + 6 = 24	

Table 3.18 : Winding Table

Phase R	1 + 13 = 14	14 + 11 = 25
	25 + 13 = 38	38 + 11 = 49 (1)
	2 + 13 = 15	15 + 11 = 26
	26 + 11 = 37	37 + 11 = 48
B	17 + 13 = 30	30 + 11 = 41
	41 + 13 = 54(6)	8 + 11 = 19
	19 + 13 = 32	32 + 11 = 43
	43 + 13 = 56 (8)	8 + 11 = 19
Y	9 + 13 = 22	22 + 11 = 33
	33 + 13 = 46	46 + 11 = 57(9)
	11 + 13 = 24	24 + 11 = 35
	35 + 13 = 48	48 + 11 = 59 (11)

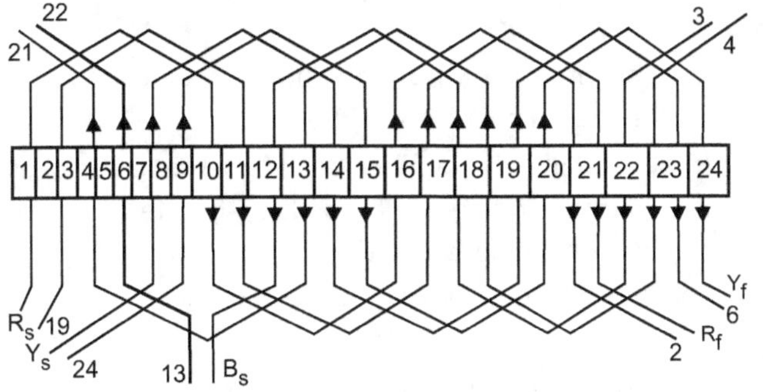

Fig. 3.38 : Winding diagram

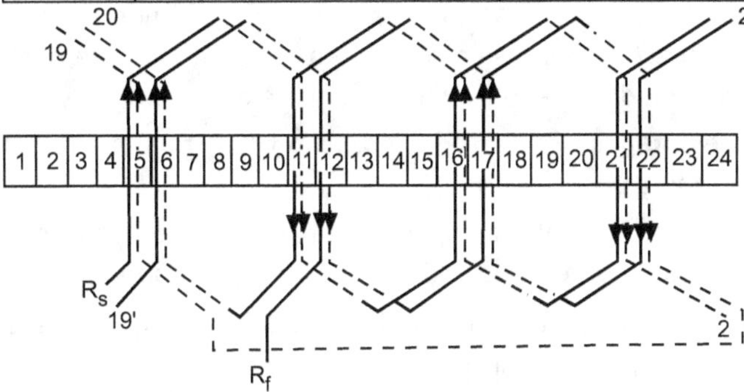

Fig. 3.39 : Winding diagram for R phase only

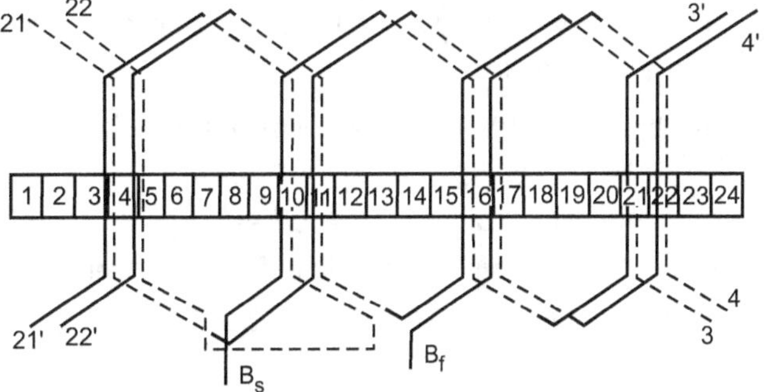

Fig. 3.40 : Winding diagram for B phase only

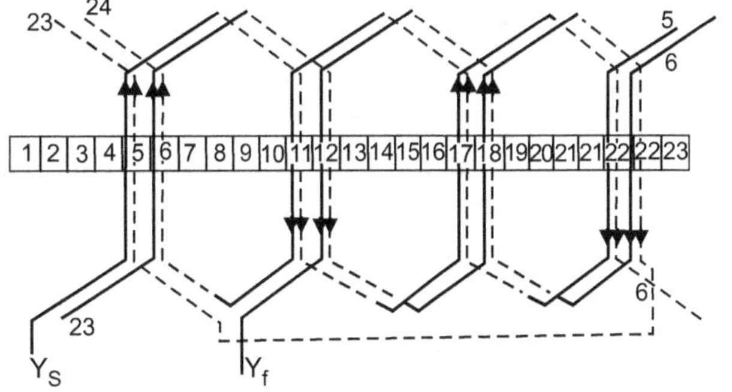

Fig. 3.41 : Winding diagram for Y phase only

Example 3.19 : *Design and draw the developed winding diagram of an AC motor with following details: No of poles = 4 no. of phases = 3, No. of slots = 24, double layer wave winding, star connected.*

Solution :

No. of poles = 4; No. of conductors = 24;

Pole pitch = 24/4 = 6; no of slots/pole /phase = 24 / (4 × 3) = 2

No. of coils = 24/2 = 12

Slot angle = 180/pole pitch = 180/6 = 30°

Winding pitch = 180 − (slot angle)

= 180 − 30 = 150

Hence coil span = 180_0 / 30 = 6 slots

Connections: R_S = 1, Y_S = 1 + 120/30 = 5; B_S = 1 + 240/30 = 9

$$Y_b = 13 \text{ and } Y_f = 11.$$

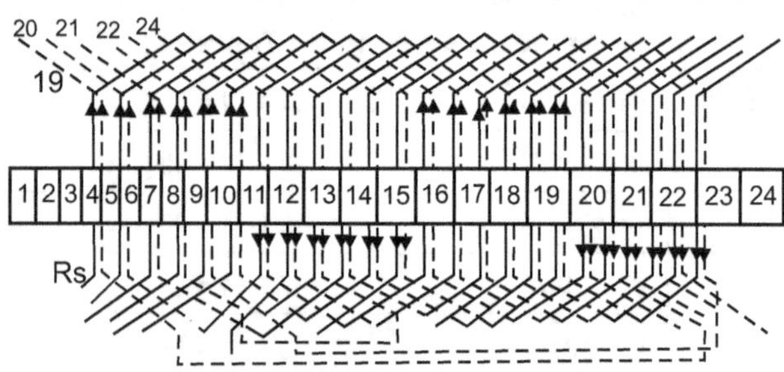

Fig. 3.42 : Winding diagram for RYB phases

3.21 SHORT CHORDED WINDING OR FRACTIONAL SLOT WINDING

- Coil pitch in poly phase machines is usually less than pole-pitch and such a winding arrangement is called short pitch or chorded or fractional slot winding.

- Usually the coil pitch varies from 2/3 pole pitch to full pole pitch.

- A coil span less than 2/3 pole pitch is not used in practice. Because a chording more than 1/3 pole pitch would noticeably reduce the phase emf.

Advantages of Short Pitched, (chorded, Fractional Slot) Windings are :

- The amount of copper used in the overhang (end winding) reduced and hence a saving on copper,

- The magnitude of certain harmonics in the emf and also mmf is suppressed.

Note :

- In integral full pitch winding, a slot contains coil sides of the same phase.

- In integral chorded pitch winding, some slots contain coil sides pertaining to different phases.

- Interconnection between the phase belts of chorded three phase winding is done in a similar manner to that explained earlier for full pitch winding.

For Example :

- Consider a motor stator with 36 slots wound for six poles.

  Such a motor will have synchronous speed of 1,000 rpm and the number of slots per pole per phase is :-

$$q = \frac{36}{6 \times 3} = 2$$

- If the same stator is rewound for the lower speed say, 750 rpm, i.e., for 8 poles, the number of slots per pole per phase will then be :

$$q = \frac{36}{8 \times 3} = 1\frac{1}{2} = \frac{3}{2}$$

- In induction motors such cases usually arise when stators with the same number of slots are wound for more than one speed or number of poles.

  For frictional slot windings, however, from the view point of symmetry, the number of slots must be divisible by the number of phase. i.e. 3

- **Limitations of Frictional Slot Windings are**

  ➢ It can be used only with double-layer windings

  ➢ The number of parallel circuits is limited

- The fractional-slot winding differs from the integral-slot winding in that it must be composed of coil groups with different numbers of coils and each phase must occupy the same number of slots, otherwise the winding would be unbalanced.

- Usually, the frictional-slot winding is a combination of two types of coil groups :

  1. One in which the number of coils in the group is equal to the integer part of the number of slots per pole per phase.

  2. The other in which the number of coils is one greater than in the first type.

- If for example, the number of slots per pole per phase is 2 ½, the winding will be built up of alternating coil groups containing two and three coils each, every two-coil group being followed by a three-coil group.

  2-3-2-3-2-3 ......

- Because of the alternation, the number of slots per pole per phase is :

- Sometimes the fractional number of slots per pole per phase is expressed as an improper fraction, i.e.

$$q = \frac{c}{d}$$

In the example above, c = 5 and d = 2

To obtain a balanced or symmetrical winding, it is necessary that $\dfrac{S}{t \cdot m}$ be equal to a whole number.

where,   S - being the number of slots,

        t – the largest common factor for S and P, and

        m – the number of phases.

Arranging Fractional Slot Windings with the Aid of Tables :

- The coil groups in a fractional-slot winding are easily arranged with the aid of a table.

- Taking a sheet of millimeter lined paper, the table is drawn with as many horizontal lines as there are poles, and each line is divided into 3C boxes, where C is the numerator of the improper fraction representing the slots per pole per phase and 3 is no. of phases.

- The table is next divided by vertical lines forming three equal columns for the three phases with C boxes per phase.

- Following this, in ordinal succession, the boxes are filled in with the numbers of the slots at intervals of d boxes, where d is the denominator of the fraction expressing the number of slots per pole per phase.

Example 3.20 : $S = 27$, $p = 6$, $m = 3$, $q = 1\frac{1}{2} = \frac{3}{2}$

Solution : The largest common factor t for $S = 27$ and $p = 6$ is :

$$S = 27 = 3 \times 3 \times 3$$

$$p = 6 = 2 \times 3$$

then, t = 3 and $S/(t/m) = 27/(3 \times 3) = 3$ is a whole number.

1. Draw a table where no. rows = no. of poles and each column of three phases with C no. of sub columns, where, C is the numerator of the improper fraction.

2. Fill the boxes from the extreme left top box with cross or consecutive numbers (representing adjacent slots) as shown in table below. Produced to the right marking crosses/numbers separated from each other by denominator of the improper fraction of no. of slots per phase per pole.

Table 3.19 : Details of Position of Conductors in Slots

No. of Poles	Phase R		Phase B		Phase Y	
N	1	2	3	4		5
S	6	7	8	9		
N	10	11	12	13		14
S	15	16	17	18		
N	19	20	21	22		23
S	24	25	26	27		

Winding Table Interpretation :

- Reading the table horizontally line by line, write down the letter of the respective phase each time a cross/number appears in its column.

- This reveals the following sequence of the coils of each phase under consecutive poles.

  RRBYY, RBBY, RRBYY, RBBY, RRBYY, RBBY.

- Each letter indicates the coils of each phase, and like letters succeeding one another indicate how many coils of the same phase the group will contain.

- Thus, in our example, the sequence shows that it is necessary to prepare nine groups of two coils each and nine single coils.

- They will occupy $(9 \times 2) + 9 = 27$ slots with the following arrangement.

$$\underline{2,1,2};\ \underline{1,2,1};\ \underline{2,1,2};\ \underline{1,2,1};\ \underline{2,1,2};\ \underline{1,2,1};$$
$$\ \ N\ \ \ \ \ \ S\ \ \ \ \ \ N\ \ \ \ \ \ S\ \ \ \ \ \ N\ \ \ \ \ \ S$$

Summary of Fractional-Slot Winding :

- When the integer before the fraction is greater than unity, the numbers in the sequence table must be that integer and a number increased by one.

- Thus, for example, when $q = 1\ \frac{1}{2}$, the sequences will contain repeating single and two-coil groups (1-2), while in the case where $q = 2\ \frac{1}{2}$ the repeating sequences will contain two-coil and three coil groups (2-3).

- The number of integers in a period is equal to the denominator d of the improper fraction expressing the slots per pole per phase; the sum of the integers is equal to c, the numerator of the improper fraction.

- Thus, when the period consists of five integers, (1-2-1-2-2), the sum of the integers is 8, i.e., it is equal to the numerator of the fraction.

Example 3.21 : *Design and draw the developed winding diagram of an AC motor with following details: No of poles = 6 no. of phases = 3, No. of slots = 27, double layer lap winding, star connected.*

Solution :

No. of poles = 6; No. of conductors = 27;

Pole pitch = 27/6 = 4.5; no of slots/pole/phase

= 27 / (6 × 3) = 1.5

Slot angle = 180/pole pitch = 180/4.5 = 40°

Winding pitch = 180 – (slot angle)

= 180 – 40 = 140

Hence coil span = 180° / 40 = 4.5 slots

Connections: Rs = 1, Ys = 1 + 120/40 = 4; Bs = 1 + 240/40 = 7

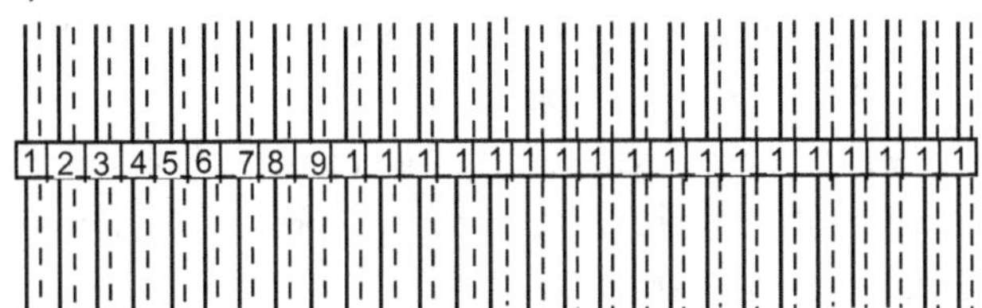

Fig. 3.43: Placement of conductors of winding diagram

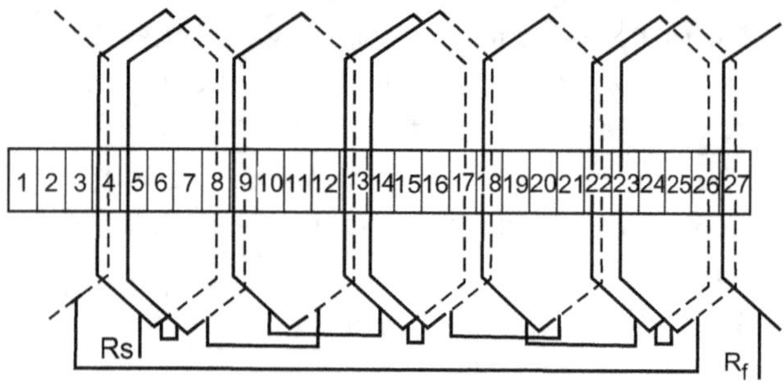

Fig. 3.44 : Winding diagram for R phase only

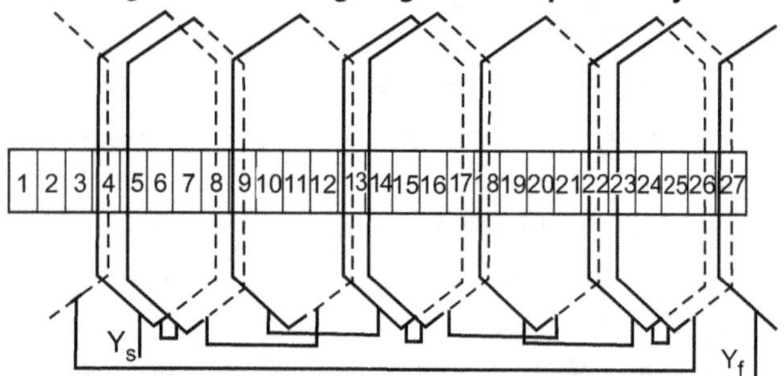

Fig. 3.45: Winding diagram for Y phase only

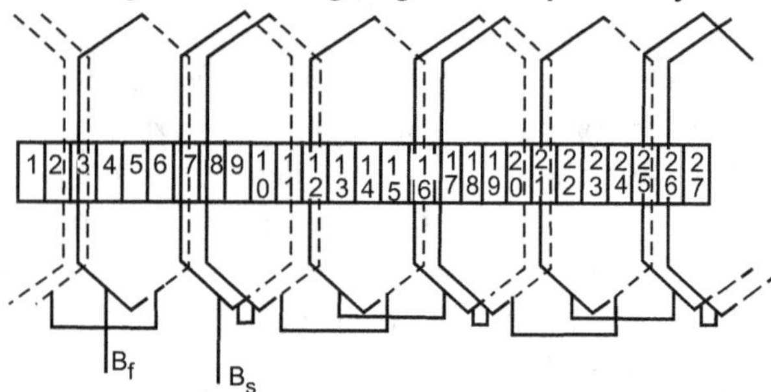

Fig. 3.46 : Winding diagram for B phase only

3.22 MUSH WINDING

- This winding is very commonly used for small induction motors having circular conductors.

- This is a single layer winding where all the coils have same span (unlike the concentric winding where coils have different spans).

- Each coil is wound on a former, making one coil side shorter than the other.

- The winding is put on the core by dropping the conductors, one by one into previously insulated slots.

- The short coil sides are placed first and then the long coil sides. The long and short coil sides occupy alternate slots.

- It will be also observed that the ends of coil situated in adjacent slots cross each other i.e. proceed to left and right alternatively.

- That is why sometimes it is known as *a basket winding*.

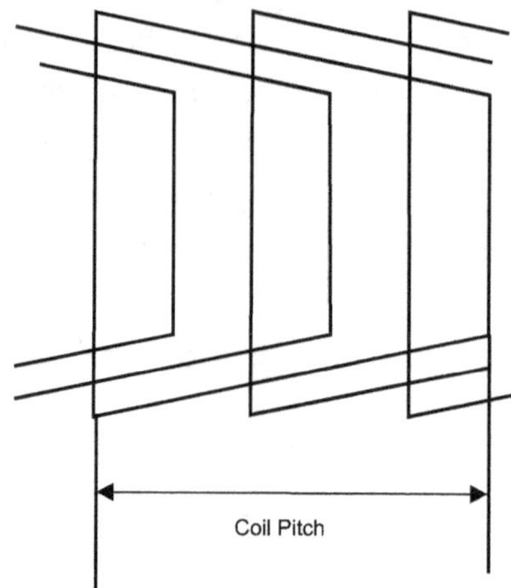

Fig. 3.47 : Mush winding over one coil pitch

Points to be Remembered:

The following should be kept in mind while designing a mush winding, that is

- The coils have a constant span.

- There is only one coil side per slot and therefore the number of coil sides are equal to number of slots.

- There is only one coil group per phase per pole pair and therefore, the maximum number of parallel paths per phase is equal to pole pair.

- The coil span should be odd. Thus for a 4 pole 36 slot machine, coil span should be 36/4=9 while for a 4 pole 24 slot machine, the coil span should not be 24/4=6; it should be either 5 or 7 slots. This is because a coil consists of a long and a short coil side. The long and short coil sides are placed in alternate slots and hence one coil will be in an even numbered slot and the other in an odd numbered slot giving a coil span which is an odd integer.

Example 3.22 : Consider the winding data with must winding

Given data : S = 12; p = 2; m = 3; type = Mush

No. of coil groups per phase = $K = 3 \cdot \dfrac{P}{2} = 3 \cdot \dfrac{2}{2} = 3$

No. of coils in a group = $q = \dfrac{S}{m \cdot p} = \dfrac{12}{3.2} = 2$

Coil pitch = $Y_S = \dfrac{S}{P} = \dfrac{12}{2} = 6$

This is an even number and hence the winding is not possible with an even coil span. Therefore, it is shortened by one slot and a coil span of 5 slots is used.

The electrical angle, $\gamma = 180 \cdot P = 180 \cdot 2 = 360°$

The angle between adjacent slots,

$$\alpha = \frac{\gamma}{S} = \frac{360°}{12} = 30°$$

The distance between the beginnings of each phase,

$$\gamma = \frac{120°}{\alpha} = \frac{120°}{30°} = 4 \text{ slots}$$

If the beginning pf Phase R is slot 1, then the beginning of phase Y is slot $1 + \lambda = 5$ and the beginning of phase B is slot $1 + 2\lambda = 1 + 8 = 9$.

Coil group of Phase A : Lay down coil-group belonging to phase A inside the slots 1, 2 and 7, 8.

Coil group of phase A : Lay down coil-group belonging to phase A inside the slots 1, 2 and 7, 8.

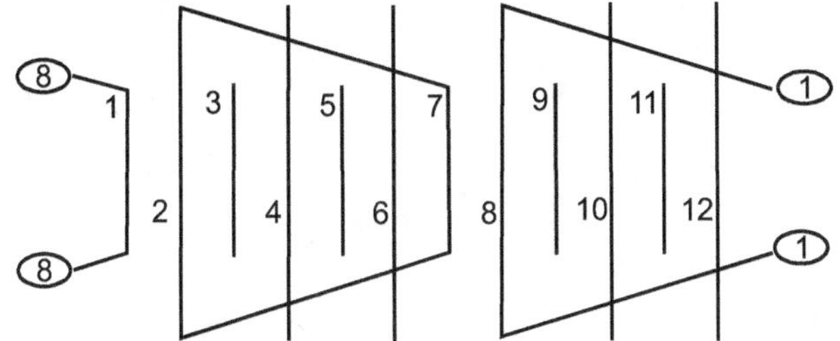

Fig. 3.48 : Winding diagram showing coil group of Phase A

Coil group of phase B :

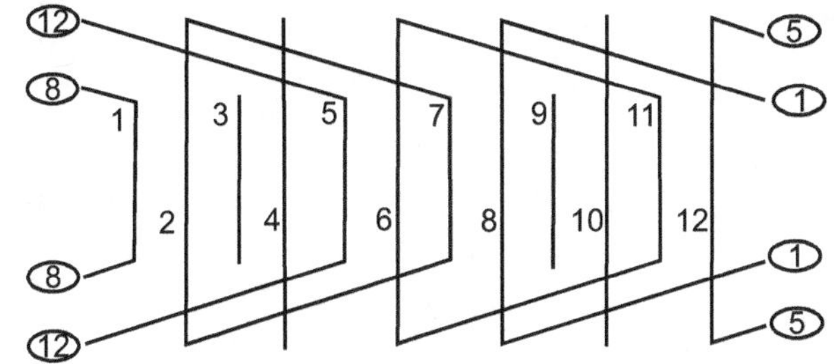

Fig. 3.49 : Winding diagram showing coil group of phase B

Coil group of Phase C :

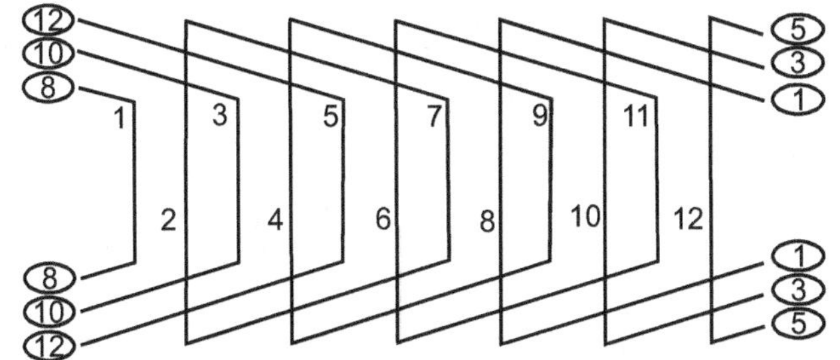

Fig. 3.50 : Winding diagram showing coil group of phase C

Phase A, B and C coil group interconnections and Terminals

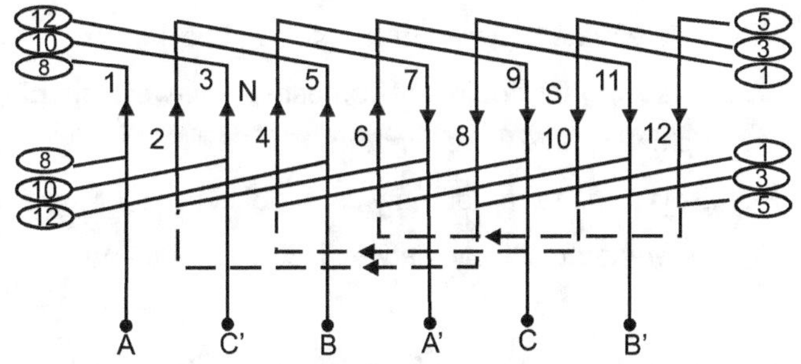

Fig. 3.51 : Winding diagram for mush winding

Example 3.23 : *Design and draw the developed winding diagram of an AC motor with following details: No of poles = 4 no. of phases = 3, No. of slots = 24, single layer mush winding.*

Solution :

No. of poles = 4; No. of conductors = 24;

Pole pitch = 24/4 = 5 (should be odd)

no of slots/pole /phase = 24 / (4 × 3) = 2

Slot angle = 180 × 2 /pole pitch = 180 × 2 / 24 = 15

Connections: $R_S = 2$, $B_S = 2 + 120/15 = 10$; $Y_S = 2 + 240/15 = 18$

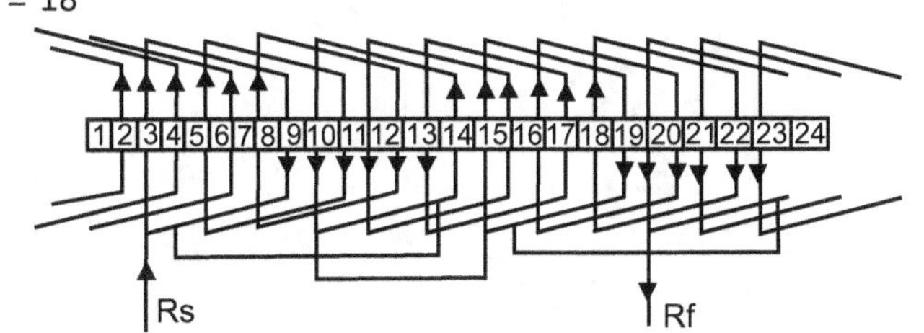

Fig. 3.52 : Winding diagram for all 3 phases

EXERCISE

1. Define the following terms

    (a) Conductor ,turn and coil side

    (b) Coil Span

    (c) Winding pitch

    (d) Commutator pitch

    (e) Full Pitch Winding

    (f) Front Pitch

2. Classify the types of windings. Make the comparison between them.

3. What is requirement of dummy a coil.

4. Explain the terms single layer and double layer windings.

5. Differentiate between closed windings and open windings.

6. Differentiate between front pitch and back pitch.

7. Give important points of comparison between simple lap and simple wave windings.

8. Explain the terms 'dummy coil' and 'equalizer rings'.

9. Differentiate clearly between 60° spread and 120° phase spread.

10. Compare : Integral slot and fractional slot winding.

11. Mention some important features of 'mush winding'.

12. Give some advantages of fractional slot winding.

13. Mention some of the advantages and disadvantages of short pitching the coil.

14. What is the effect, of chording and distributing the winding in slots, in the emf generated.

15. What is role of equalizer connection.

◈ ◈ ◈

HEATING, COOLING AND VENTILATION

4.1 INTRODUCTION

- During normal operation of any electrical machine, certain losses take place in various parts of the machine, which ultimately raise the temperature of those parts. Before dealing with a detailed study of any electrical machine, it is necessary to discuss the causes of temperature rise in electrical machines and various methods to dissipate these losses, so that the temperature rise of any part of machine does not exceed the permissible limit, imposed by the insulating materials used in that machine.

- The total losses occurring in an electrical machine is converted into heat and as a result, the various parts of the machine are heated and their temperature rises above the ambient temperature. In order to ensure reliable operation of electrical machines, the temperature rise of any part at rated load should not be higher than the limiting temperature.

- The various parts of the machine dissipate heat by radiation, convection and conduction. In static electrical machines, e.g. transformers, the heat dissipated by conduction is negligible, whereas heat dissipated by convection is maximum. Thus in static electrical machines, heat is dissipated mainly by convection and radiation. In rotating electrical machines, heat is dissipated mainly by convection and radiation. In rotating electrical machines, heat is dissipated mainly by a combination of convection and conduction. Heat dissipated by radiation from the outside surface is comparatively low in case of electrical machines with moving parts.

4.2 MODES OF HEAT DISSIPATION

- The process of energy transfer in the case of transformers and electro-mechanical energy conversion in the case of rotating electrical machines involves currents in the conductors, and fluxes in the ferromagnetic parts. Thus there are I^2R losses in windings and core losses in the ferromagnetic cores. In addition losses occur in tank walls, end plates and covers on account of leakage flux. The losses appear as heat and therefore the temperature of every affected part of the machine rises above the ambient medium which is normally the surrounding air. The heated parts of an electrical machine dissipate heat into their surroundings by conduction and convection assisted by radiation.

4.2.1 Conduction

This mode of dissipation of heat is important in the case of solid parts of machine like copper, iron and insulation. Consider two points in an electric circuit potentials V_1 and V_2, the current flowing between them is

$$I = \frac{(V_1 - V_2)}{R}$$

where R is the electrical resistance of the conducting medium between them. Similarly, we can write the equation for heat flow for conduction between two surfaces separated by a heat conducting medium, as :

$$Q_{con} = \frac{\theta_1 - \theta_2}{R} \qquad \ldots (4.1)$$

where Q_{con} = heat dissipated by conduction, W;

θ_1, θ_2 = temperatures of two bounding surfaces, °C;

R_θ = thermal resistance of the conducting medium, thermal ohm (or in °C/W).

Thermal Resistance : The thermal resistance is defined as the thermal resistance which causes a drop of 1°C per watt of heat flow. Eq. 4.1 permits heat conduction problems to be solved by methods of calculation similar to those used in electric circuits. The thermal resistance, like electrical resistance, can be written as :

$$R_\theta = \frac{\rho t}{S} \qquad \ldots (4.2)$$

$$= \frac{t}{\sigma S} \qquad \ldots (4.3)$$

ρ = thermal resistivity of material, Ω (thermal) m or °C-m/W ;

σ = $1/\rho$ = thermal conductivity, W/°C – m ;

t = length of medium, m ;

S = area of surface separated by the mediums, m^2.

Eq. 4.1 can be written as :

$$Q_{con} = \frac{S(\theta_1 - \theta_2)}{\rho t} \text{ W} \qquad \ldots (4.4)$$

Heat dissipated per unit surface area by conduction is :

$$Q_{con} = \frac{\theta_1 - \theta_2}{\rho t} \text{ W/m}^2 \qquad \text{... (4.5)}$$

The temperature difference across the conducting medium

$$\theta = \theta_1 - \theta_2$$

$$= Q_{con} R_c \qquad \text{... (4.6)}$$

$$= Q_{con} \left(\frac{\rho t}{S}\right) \qquad \text{... (4.7)}$$

Considering Eq. 4.7, we find that a material having a high value of thermal resistivity will dissipate less among of heat or alternatively for the dissipation of same heat the temperature rise will be higher.

Table 4.1 gives the values of thermal sensitivities of different materials.

Table 4.1 : Thermal Resistivities

Material	Thermal Resistivity (ohm metre)	Material	Thermal Resistivity (ohm metre)
Air (Still)	20	Asbestos	4
Cotton Cloth	14	Empire cloth	
Micanite	8	Mica	3
Compressed paper	8	Sheet steel	
Paper	7.5	Along laminations	0.02
Transformer coil	6.25	Across laminations	0.05 to 0.1
Pressboard	6	Brass	0.01
Varnished cloth	5	Aluminium	0.005
Mica tape	2.6 to 6.6	Copper	0.0026

From the Table 4.1, we have $\rho = 20$ for air and 7.5 for paper. Air has a greater thermal resistivity than paper and thus the presence of air pockets in the insulation of a machine would have disastrous effects on heat dissipation, resulting in large temperature rises.

Let us consider the case of heated where the coolant (cooling medium) is a fluid and takes away the heat by conduction. Perfect contact between the heated surface and the coolant at the outer face is rare. Therefore, the temperature difference θ is mainly dependent upon the fluid flow conditions i.e. whether the flow is stream line or turbulent and upon the condition of the surface. There may be interfacing oxide films, gas bubbles in liquids, and a stagnant surface layer of fluid and therefore, the thermal conductivity of coolant is much smaller than that of metals. The heat conducted across the interface is :

$$q_{con} = \rho_c(\theta_s - \theta_f) \text{ W/m}^2 \qquad \text{... (4.8)}$$

where θ_s and θ_f are the temperatures of surface and fluid respectively and ρ_c has a value of 50 – 1500 for oil and 10 – 50 for air.

SOLVED EXAMPLES

Example 4.1 : *A copper bar 12 mm in diameter is insulated with micanite tube which fits tightly around the bar and into the rotor slot of an induction motor. The micanite tube is 1.5 mm thick and its thermal resistivity is 8 Ωm. Calculate the loss that will pass from copper bar to iron if a temperature difference of 25 °C is maintained between them. The length of bar is 0.2 m.*

Solution : Consider Fig. 4.1.

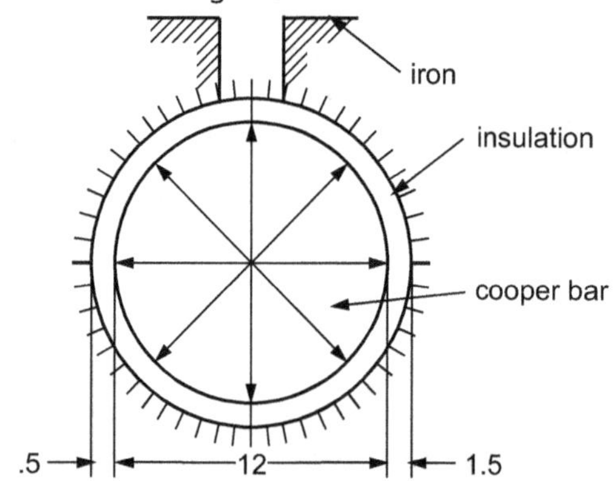

Fig. 4.1 : Conduction from copper bar
(all dimensions in mm)

Area of insulation in the path of heat flow

$$S = \text{periphery of micanite at mean radius} \times \text{length}$$

$$= \pi(12 + 15) \times 10^{-3} \times 0.2$$

$$= 4.48 \times 10^{-3} \text{ m}^2$$

From eq. 4.2, thermal resistance of micanite tube,

$$R_\theta = \frac{\rho t}{S} = \frac{8 \times 1.5 \times 10^{-3}}{8.48 \times 10^{-3}} = 1.415 \ \Omega$$

∴ From Eq. 4.1 heat dissipated,

$$Q_{con} = \frac{\theta_1 - \theta_2}{R_\theta}$$

$$= \frac{25}{1.415} = 17.67 \text{ W.}$$

Example 4.2 : *The thermal conductivity of assembled armature laminations is 20 times as great along the direction of laminations as in the direction across the laminations. Calculate the loss that will be conducted across the laminations in a stack 40 mm thick and 6000 mm^2 in cross-section with a difference of 20°C. Given that a difference of 5°C will cause 25 W to be conducted through a cross-section of 2500 mm^2 in area and 20 mm thick measured along the laminations.*

Solution : Given :

$$Q_{con} = 25 \text{ W}, t = 20 \text{ mm} = 20 \times 10^{-3} \text{ m}$$
$$S = 2500 \text{ mm}^2 = 25 \times 10^{-4} \text{ m}^2, \text{and}$$
$$\theta = \theta_1 - \theta_2 = 5°C.$$

This data refers to a direction along the laminations.

From Eq. 4.4,

$$Q_{con} = \frac{S(\theta_1 - \theta_2)}{\rho t} \text{ or } 25 = \frac{25 \times 10^{-4} \times 5}{\rho \times 2 \times 10^{-2}}$$

$\therefore$ Thermal resistivity along the direction of laminations

$$\rho = 0.025 \text{ }\Omega m.$$

Hence thermal resistivity across the laminations

$$= 20 \times 0.025 = 0.5 \text{ }\Omega m.$$

$$\left(\text{It is given that ratio } \frac{\text{resistivity across the laminations}}{\text{resistivity along the laminations}} = 20 \right)$$

Applying Eq. 4.4, heat conducted across the laminations

$$Q_{con} = \frac{S(\theta_1 - \theta_2)}{\rho t} = \frac{6000 \times 10^{-6} \times 20}{0.5 \times 40 \times 10^{-3}} = 6W$$

4.2.2 Radiation

The heat dissipated by radiation from a surface depends upon its temperature and its other characteristics like colour, roughness etc.

For the case of a very small spherical radiating surface inside a large and or black spherical shell, the heat radiated per unit surface is given by Stefan Boltzmann law :

$$q_{rad} = 5.7 \times 10^{-8} \, e\left(T_1^4 - T_0^4\right) \text{ W/m}^2 \qquad \dots (4.9)$$

where, T_1, T_0 = absolute temperatures of the emitting surface and the ambient medium respectively, K;

θ_1, θ_0 = temperatures of the emitting surface and the ambient medium respectively, °C;

e = co-efficient of emissivity ;

= 1 for perfect black bodies and is always less than unity for others.

Table 4.2 gives the co-efficient of emissivity and absorption factor for different types of surfaces.

Table 4.2 : Emissivity and Absorption Factors

Surface	Emissivity e	Absorption Factor e_a
Aluminium	0.10	0.15
Copper	0.15	–
Steel		
Rough	0.24	–
Sheet	0.55	–
Metal Paint Aluminium	0.55	0.55
Lead Paints		
White	0.90	0.25
Grey	0.95	0.75

Eq. 4.9 can be written as :

$$q_{rad} = 5.7 \times 10^{-8} \, e(T_1 - T_0)$$
$$\left[T_1^3 + T_1^2 T_0 + T_1 T_0^2 + T_0^3\right] \qquad \dots (4.10)$$

Now $(T_1 - T_0) = [(273 + \theta_1) - (273 + \theta_0)] = \theta_1 - \theta_0 = \theta$ represents the temperature rise the body, above the ambient medium.

The term $\left(T_1^3 + T_1^2 T_0 + T_1 T_0^2 + T_0^3\right)$ varies relatively little within conventional temperature limits for electrical machines and so Eqn. 4.9 can be written as :

$$q_{rad} = 5.7 \times 10^{-8} \, eK_r(T_1 - T_0)$$
$$= 5.7 \times 10^{-8} \, eK_r(\theta_1 - \theta_0) \text{ W/m}^2$$

where, $K_r = \left[T_1^3 + T_1^2 T_0 + T_1 T_0^2 + T_0^3\right].$

The value of 'e' for cast iron or steel surface, varnished insulation etc. is between 0.97. Assuming a value of 0.83 for safety, we have :

$$q_{rad} = 5.7 \times 10^{-8} \times 0.83 \, K_r(\theta_1 - \theta_0)$$
$$= 4.8 \times 10^{-8} \, K_r(\theta_1 - \theta_0) = \lambda_{rad}(\theta_1 - \theta_0)$$
$$= \theta_{rad} \cdot \theta \text{ W/m}^2 \qquad \dots (4.11)$$

where, $\lambda_{rad} = 4.8 \times 10^{-8} \, K_r$

= modified specified heat dissipation or emissivity measured in W/m^2 at a temperature rise of 1°C. Its value depends upon the temperature θ_1 and θ_0 and hence is not constant.

Table 4.3 gives the value of λ_{rad} with respect to temperature rises and temperatures of ambient medium.

Table 4.3 : Specific Heat Dissipation (Radiation) in W/m^2 – °C

Temperature Rise θ °C	Temperature of Ambient Medium θ_0 °C	
	20	30
5	5.03	5.55
20	5.42	5.98
40	6.05	6.59
80	7.30	7.98

The total heat dissipated by radiation is :

$$Q_{rad} = q_{rad} \times S = \lambda_{rad}\, \theta\, S \text{ watt} \qquad ... (4.12)$$

In electrical rotating machines and transformers radiation does not normally occur by itself and in almost every case it is accompanied by convection. Therefore, the following expression may be conveniently used,

$$q_{rad} = 2.9 \text{ e}\theta^{1.17} \text{ W/m}^2 \qquad ... (4.13)$$

Examining Table 4.2, we find that the value of co-efficient of emissivity for dull metallic paint is 0.9 while for polished metal it is 0.15. This means that the specific heat dissipation (W/m^2 – °C) due to radiation for surfaces painted with dull metallic paints is large. Hence all the electrical machines are painted with dull metallic paints (usually grey in color) in order to have a large heat dissipation due to radiation. This keeps the temperature rise of the machines to a low value.

Some electrical apparatus, particularly transformers are used fir outdoor duty. They may absorb radiant heat from the sun by Insolation. The earth's outer atmosphere receives about 1.3 kW/m^2 and under favorable climatic and air conditions, about two third of this energy may each the earth's surface. In case the absorption factor of a surface is high, (approaching unity) it may absorb large energy by insolation thereby its temperature rises by a few degrees. However, if the absorption factor is small, the sun's radiation is re-emitted.

Example 4.3 : *A heat radiating body can be assumed to be spherical surface with co-efficient of emissivity = 0.8. The temperature of the body is 60 °C and that of the walls of the room, in which it is placed, is 20 °C. Find the heat radiated from the body in watt per square metre.*

Solution :

We have $\quad e = 0.8, T_1 = 273 + 60 = 333$ °K

$$T_0 = 273 + 20 = 293 \text{ °K}$$

Using Eq. 4.9

Heat radiated

$$q_{rad} = 5.7 \times 10^{-8} \times 0.8\,(333^4 - 293^4)$$
$$= 224.6 \text{ W/m}^2$$

Example 4.4 : *A 250 V, 1 kW, single element resistor is made from 0.2 mm thick nickel chrome strip. The temperature rise of strip is not to exceed 300 °C over the ambient temperature of 30 °C. Calculate the length and width of the strip. Assume, emissivity = 0.9 radiating efficiency = 0.75, resistivity of nickel chrome 1 $\times 10^{-6}$ Ωm.*

Solution : Temperature of strip

$$\theta_1 = 300 + 30 = 330°C$$

Temperature of ambient medium

$$\theta_0 = 30°C.$$

Absolute temperature of strip

$$T_1 = 330 + 273 = 603 \text{ K}$$

Absolute temperature of ambient medium

$$T_0 = 30 + 273 = 303 \text{ K}$$

Co-efficient of emissivity = 0.9

Effective co-efficient of emissivity,

$$e = 0.75 \times 0.9 = 0.675.$$

From Eq. 4.9, heat dissipated by radiation

$$q_{rad} = 5.7 \times 10^{-8} \text{ e}\,(T_1^4 - T_0^4)$$
$$= 5.7 \times 10^{-8} \times 0.675\,(603^4 - 303^4)$$
$$= 4760 \text{ W/m}^2$$

Resistance of strip

$$R = \frac{\text{voltage}^2}{\text{power}} = \frac{(250)^2}{1000} = 62.5 \ \Omega$$

Let, $\quad l = $ length of strip, m;

$\qquad w = $ width of strip, m;

and $\quad t = $ thickness of strip, m.

Area of cross-section of strip

$$= w \times t = w \times 0.2 \times 10^{-3}$$
$$= 0.2w \times 10^{-3} \text{ m}^2$$

Resistance of strip

$$R = \frac{\rho l}{wt}$$

$$= \frac{1 \times 10^{-6}\, l}{0.2 \times 10^{-3}\, w} = \frac{5l}{w} \times 10^{-3} \ \Omega$$

$\therefore (5l/w) \times 10^{-3} = 62.5$ or $l/w = 12500$

Heating dissipating surface of strip, neglecting edges,

$$S = 2\, lw$$

Total heat radiated = $q_{rad} \times S = 4760 \times 2lw = 9520\, lw$

$\therefore \quad 9520\, lw = 1000$ or $lw = 0.105$

But $\quad l/w = 12500 \therefore l^2 = 0.105 \times 12500 = 1312.5$

or $\quad l = 36.2$ m

and $\quad w = \dfrac{0.105}{36.2}$ m $= 2.9$ m

4.2.3 Convection

Heat dissipation by convection is classified into two categories :

1. Natural and
2. Artificial

1. **Natural Convection :** Liquid and gas particles near heated body becomes lighter and rise, giving place to cooler particles which in turn get heated and rise. This natural process, due to change in fluid density is known as natural convection.

The heat dissipated per unit surface by natural convection is given by

$$q_{conv} = K_c (\theta_1 - \theta_0)^n \text{ W/m}^2 \qquad \dots (4.14)$$

where K_c = a constant depending on the shape and dimensions of hot body

n = a constant depending upon shape and dimensions of hot body. Its value lies between 1 and 1.25.

θ_1 = temperature of emitting surface, °C;

θ_0 = temperature of ambient medium, °C;

Taking $n = 1$

Heat dissipated per unit area by convection,

$$q_{conv} = K_c (\theta_1 - \theta_0) = \lambda_{conv} \theta \qquad \dots (4.15)$$

where λ_{conv} = specific heat dissipation or emissivity due to convection measured in $\text{W/m}^2 - °C$.

Total heat dissipated by convection,

$$Q'_{conv} = q_{conv} S = \lambda_{conv} \theta S \text{ watt} \qquad \dots (4.16)$$

Convection is a complicated phenomenon and heat convected depends upon many variables such as

(i) Power density

(ii) Temperature difference between heated surface and coolant

(iii) Height, orientation, configuration and condition of heated surface

(iv) Thermal resistivity, density, specific heat, viscosity and co-efficient of volumetric expansion of fluid, and

(v) Gravitational constant.

In order to find the heat convected the following formulae have been developed for the temperature range and structure of machines and transformers. For vertical planes of height not less than 1 metre and with a temperature difference θ between the surface and the ambient air of pressure p atmosphere,

$$q_{conv} = 2 \, p^{0.5} \, \theta^{1.25} \text{ W/m}^2 \qquad \dots (4.17)$$

For vertical tubes of diameter d metre,

$$q_{conv} = 1.3 d^{-0.25} \, \theta^{1.25} \text{ W/m}^2 \qquad \dots (4.18)$$

2. **Artificial Convection :** In modern machines heat is removed by artificial circulation of cooling medium. For example, a transformer tank may be cooled buy blasting air on it or a turbo-alternator may be cooled by circulating hydrogen. This is known as cooling by artificial convection. The usual method employed for cooling of machines is by blasting air on heating surfaces; there surfaces may be open or closed. The increase in heat dissipation by air blasts is due to increase in convection. The problem of calculation of heat dissipation by artificial convection is very complex as it mainly depends on the constructional features of the machine. These constructional details are different for every machine and so no exact relationship can be given for artificial convection. However, one of the most widely used formulae for air blasts on open surfaces is :

$$\lambda'_{conv} = \lambda_{conv} (1 + K_v \sqrt{V}) \text{ W/m}^2 - °C \qquad \dots (4.19)$$

where, λ'_{conv} = specific heat dissipation of a blasted surface.

λ_{conv} = specific heat dissipation by natural convection.

V = relative velocity of cooled surface and air blast, m/s

K_v = a constant, depending upon whether the blast is uniform or non-uniform,

= 1.3 for uniform blasts.

The value of K_v comes down to even 0.5 for non-uniform blasts.

4.3 TEMPERATURE RISE – TIME RELATION

- The temperature of a machine rises when it is run under steady load conditions starting from cold conditions. The temperature at first increases at a rate determined by power wasted. As the temperature rises, the active parts of the machine dissipate heat partly by conduction, partly by radiation, and in most cases, largely by means of air cooling. The higher the temperature rise, the greater would be the effect of these methods of cooling. Therefore, as the temperature rises, its rate of increase falls off owning to better heat dissipating conditions. As shown later, the temperature-time curve is exponential in nature.

- The temperature of any part of a machine, not only depends on the heat produced in itself but also on heat produced in other parts. This is because there is always a heat flow from one part to another ; for example, the heat produced in the part of the winding embedded in the slot flows partially through the insulation to the laminations partially to the end windings. Thus the end windings have to transfer to the air, not only the heat produced in them but also a part of the heat produced in the slot portion of the winding.

- Electrical machines are not homogeneous bodies. Their parts are made up of different materials like copper, iron and insulation. These materials have different thermal resistivities and due to this, it is rather difficult to calculate the temperature of a part of a machine. However, it is worthwhile taking theory of heating of homogeneous bodies as the basis for analysing the process of machine heating. The results obtained from such a theory are applicable to a certain degree, to the different parts of machine as a whole.

Let, Q = power loss or heat developed, J/s or W;

G = weight of active parts of machine, kg;

h = specific heat, J/kg-°C;

S = cooling surface, m^2;

λ = specific heat dissipation, W/m^2-°C;

c = $1/\lambda$ = cooling co-efficient, °C-m^2/W;

θ = temperature rise at any time t, °C;

θ_m = final steady temperature rise while heating, °C;

θ_n = final steady temperature rise while cooling, °C;

θ_i = initial temperature rise over ambient medium, °C;

T_h = heating time constant, s;

T_c = cooling time constant, s;

t = time, s.

4.3.1 Heating

Considering the conditions at any time t from start, heat energy developed in the body during an infinity small time dt,

= heat energy developed per second $\times$ dt = Qdt ... (4.19)

If during this period dt the temperature of the body rises by dθ, the heat energy stored in the body,

= weight of body $\times$ specific heat
$\times$ difference in temperature

= G h dθ ... (4.20)

If in the process of heating, the temperature of the surface rises by θ over the ambient medium, at the instant considered, the heat energy dissipated by the body into the ambient medium due to radiation, conduction and convection,

= specific heat dissipation $\times$ surface
$\times$ temperature rise $\times$ time

= $\lambda \times S \times \theta \times dt = S \lambda \theta$ dt.   ... (4.21)

As the heat developed in the machine is equal to the heat stored in the parts plus the heat dissipated, we have from Eqs. (4.19, 4.20 and 4.21,

$$Q\,dt = G\,h\,d\theta + S\lambda\theta\,dt \qquad \text{... (4.22)}$$

or

$$dt = \frac{d\theta}{\dfrac{Q}{Gh} - \dfrac{S\lambda}{Gh}\theta} \qquad \text{... (4.23)}$$

Solving the differential equation 4.23,

$$t = \frac{Gh}{S\lambda}\log_e\left(\frac{Q}{Gh} - \frac{S\lambda}{Gh}\theta\right) + K \qquad \text{... (4.24)}$$

where K is the constant of integration.

The value of K is found by applying the boundary condition,

when t = 0, we have $\theta = \theta_i$.

we have :

$$0 = -\frac{Gh}{S\lambda}\log_e\left(\frac{Q}{Gh} - \frac{S\lambda}{Gh}\theta\right) + K$$

or

$$K = -\frac{Gh}{S\lambda}\log_e\left(\frac{Q}{Gh} - \frac{S\lambda}{Gh}\theta_i\right).$$

Substituting this value of K in eq. 4.24

$$t = -\frac{Gh}{S\lambda}\log_e\left(\frac{Q}{Gh} - \frac{S\lambda}{Gh}\theta\right) + \frac{Gh}{S\lambda}\log_e\left(\frac{Q}{Gh} - \frac{S\lambda}{Gh}\theta_i\right)$$

$$= -\frac{Gh}{S\lambda}\log_e\frac{\dfrac{Q}{S\lambda} - \theta}{\dfrac{Q}{S\lambda} - \theta_i} \qquad \text{... (4.25)}$$

The machine reaches a final steady temperature rise θm when t = ∞. Under this condition there is no further temperature rise and the rates of heat production and dissipation are equal. This means

dθ = 0 or Ghdθ = 0

when the machine attains final steady temperature rise.

From Eq. 4.22,

for $\theta = \theta_m$,

$$Q_{dt} = S\lambda\theta_m\,dt$$

or $\theta_m = Q/S\lambda$ ... (4.26)

$$= Q_c/S \qquad \text{... (4.27)}$$

Putting $\theta_m = \dfrac{Q}{S\lambda}$ in eq. 4.25, we have

$$t = \frac{Gh}{S\lambda} \log_e \frac{\theta_m - \theta}{\theta_m - \theta_i} \qquad \text{... (4.28)}$$

The term $Gh/S\lambda$ has the dimensions of time and is called the heating time constant T_h.

$$\therefore \qquad T_h = Gh/S\lambda \qquad \text{... (4.29)}$$

Putting this value of

$$T_h = \frac{Gh}{S\lambda} \text{ in eq. 4.28}$$

$$t = T_h \log_e \left[\frac{\theta_m - \theta}{\theta_m - \theta_i} \right] \qquad \text{... (4.30)}$$

or $\quad \dfrac{\theta_m - \theta}{\theta_m - \theta_i} = r^{-t/T_h}$

or $\qquad \theta = \theta_m (1 - e^{-t/T_h}) + \theta_i\, e^{-t/T_h} \qquad \text{... (4.31)}$

If the machine starts from cold conditions,

$$\theta_i = 0. \text{ (No temperature rise over the}$$
$$\text{ambient medium)}$$

$$\theta = \theta_m (1 - e^{-t/T_h}) \qquad \text{... (4.32)}$$

Relation 4.32 is the equation of temperature rise with time. The temperature rise-time curve is exponential in nature as shown in Fig. 4.2.

$$\theta = \theta_m (1 - e^{-t/T_h}).$$

Putting $\quad t = T_h$ in the above expression, we have

$$\theta = \theta_m (1 - e^{-1}) = 0.632\, \theta_m.$$

Thus we can define the heating time as the time taken by the machine to attain 0.632 of its final steady temperature rise. The heating time constant of a machine is the index of time taken by the machine to attain its final steady temperature rise.

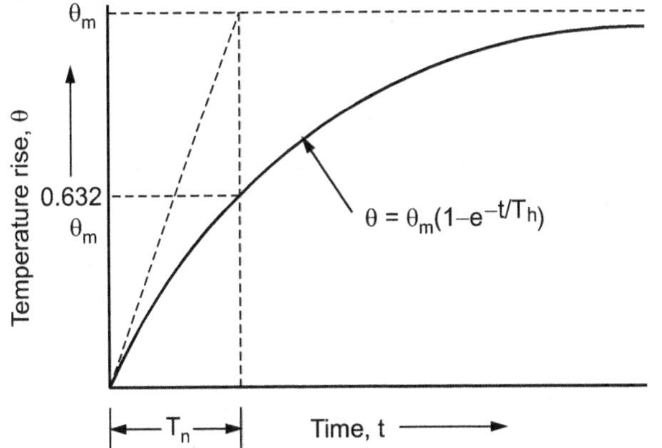

Fig. 4.2 : Heating curve

Considering the relationship $T_h = Gh/S\lambda$ we conclude that the time constant is inversely proportional to λ (specific heat dissipation). λ has a large value for well ventilated machines and thus the value of their heating time constant

is small. The value of heating time constant is large for poorly ventilated machine.

Since the volume of machine and hence its weight increases in proportion to the third power, and the surface area in propagation to second power of its linear dimensions, the heating time constant of a machine increase as first power of linear dimensions. Because of this, large sized machines have large heating time constants.

The heating time constant of a well-ventilated induction motor of about 20 kW rating may be of the order of minutes while that of large or totally enclosed machines may reach several hours or even days. The heating time constant of conventional electrical machines is usually within the range of $\dfrac{1}{2}$ to 3-4 hours.

Final Steady Temperature Rise : Considering the expression $\theta_m = Q/S\lambda$, it is clear, other things being equal, the final steady temperature rise is directly proportional to the losses. It is also evident that the final steady temperature rise is inversely proportional to surface area and specific heat dissipation. Thus for the same loss, the machine would attain a higher temperature rise if its dissipating surface is small or, if its ventilation is poor.

4.3.2 Cooling

The equation for cooling can be derived by considering eq. 4.22.

$$Qdt = Ghd\theta + S\lambda\theta\, dt$$

The solution of above equation is :

$$t = -\frac{Gh}{S\lambda} \log_e \left[\frac{Q}{Gh} - \frac{S\lambda}{Gh}\theta \right] + K$$

The value of K is obtained by putting boundary conditions,

when $\qquad t = 0, \theta = \theta_i$. From this we have,

$$0 = -\frac{Gh}{S\lambda} \log_e \left[\frac{Q}{Gh} - \frac{S\lambda}{Gh}\theta_i \right] + K$$

or $\qquad K = \dfrac{Gh}{S\lambda} \log_e \left[\dfrac{Q}{Gh} - \dfrac{S\lambda}{Gh}\theta_i \right]$

Substituting this value of K and proceeding as in the case of heating, we get,

$$\theta = \theta_n (1 - e^{-t/T_e}) + \theta_i e^{-t/T_e} \qquad \text{... (4.33)}$$

where $\qquad \theta_n = Q/S\lambda \qquad\qquad \text{... (4.34)}$

$$= Qc/S \qquad\qquad \text{... (4.35)}$$

and $\qquad T_c = Gh/S\lambda \qquad\qquad \text{... (4.36)}$

The value of λ under cooling conditions is usually different from that under heating conditions and so the heating and cooling time constants of a machine may have different values.

If a machine is shut down, no heat is produced and so its steady temperature rises when cooling is zero or $\theta_n = 0$. Under these conditions Eq. 4.33 thus reduces to

$$\theta = \theta_i e^{-t/T_e} \qquad \qquad ...\ (4.37)$$

It is clear from Eq. 4.37, that the cooling curve is also exponential in nature as shown in Fig. 4.3. Eq. 4.37 is applicable is machines which are shut down while Eq. 4.33 is applied to machines allowed to cool owing to partial removal of load.

Cooling Time Constant : Consider the relation,

$$\theta = \theta_i\, e^{-t/T_e}$$

Putting, $\quad t = T_c$, we have

$$\theta = \theta_i\, e^{-1} = 0.368\ \theta_i.$$

Thus we can define the cooling time constant as the time taken by the machine for its temperature rise to fall to 0.368 of its initial value.

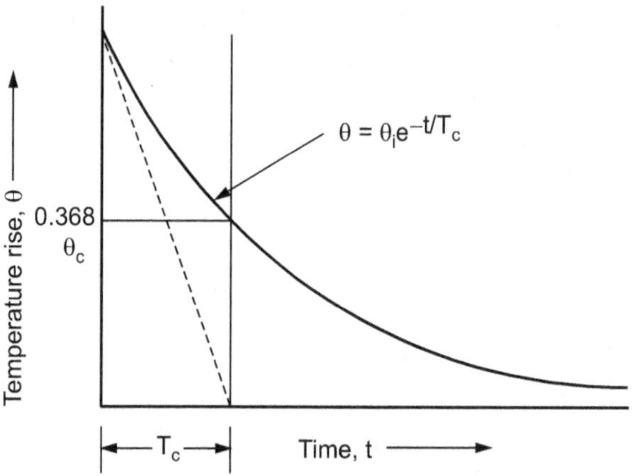

Fig. 4.3 : Cooling curve

As stated earlier the cooling time constant may be different from heating time constant for the same machine, as the ventilation conditions in the two cases may not be the same. The cooling time constant is usually larger owing to poorer ventilation conditions when the machine cools. In self cooled motors the cooling time constant is about 2-3 times greater than the heating time constant because cooling conditions are worse at stand still.

Final Steady Temperature Rise : Considering the relation $\theta_n = Q/S\lambda$, the final steady temperature rise when the machine is cooling is directly proportional to losses. When the machine is shut down $Q = 0$ and so the final temperature rise is zero, i.e. the machine finally comes to ambient temperature conditions.

Example 4.5 : *A field coil has a heat dissipating surface of $0.15\ m^2$ and a length of mean turn of 1 m. It dissipates loss of 150W, the emissivity being 34 $W/m^2 - °C$. Estimate the final steady temperature rise of the coil and its time constant if the cross-section of the coil is $100 \times 50\ mm^2$. Specific heat of copper is 390 J/kg - °C. The space factor is 0.56. Copper weights 8900 kg/m^3.*

Solution :

Volume of copper $= 1 \times 100 \times 50 \times 0.56 \times 10^{-6}$

$$= 2.8 \times 10^{-3}\ m^3.$$

weight of copper

$$G = 2.8 \times 10^{-3} \times 8900 = 24.92\ kg$$

Final steady temperature rise, from Eq. 4.26,

$$\theta_m = \frac{Q}{S\lambda} = \frac{150}{0.15 \times 34} = 29.4°C$$

Heating time constant, from Eq. 4.48,

$$T_h = \frac{Gh}{S\lambda} = \frac{24.92 \times 390}{0.15 \times 3.4} = 1906\ s.$$

Example 4.6 : *The exciting coil of an electromagnet has a cross-section of $120 \times 50\ mm^2$ and a length of mean turn of 0.8 m. It dissipates 150 W continuously. Its cooling surface is $0.125\ m^2$ and specific heat dissipation is 30 W/m^2-°C. Calculate the final steady temperature rise of the coil surface. Also calculate the hot spot temperature rise of the coil if the thermal resistivity of insulating material used is 8 Ωm. The space factor is 0.56.*

Solution : From eq. 4.45, final steady temperature rise of coil surface

$$\theta_m = \frac{Q}{S\lambda} = \frac{150}{0.125 \times 30} = 40°C$$

From Eq., effective thermal resistivity

$$\rho_o = \rho_i \left(1 - S_f^{1/2}\right) = 8(1 - 0.562^{1/2}) = 2\ \Omega\ m.$$

Loss $\quad q = \dfrac{150}{120 \times 50 \times 10^{-6} \times 0.8} = 31250\ W/m^3$

From eq. 4.23, temperature difference between coil surface and hot spot

$$\theta_0 = \frac{q\rho t^2}{8} = \frac{31250 \times 2 \times (50 \times 10^{-3})^2}{8} = 19.5°C$$

$\therefore\quad$ temperature rise of hot spot $= 40 + 19.5 = 59.5\ °C$

Example 4.7 : *A 1000 A shunt consists of 8 strips of nickel-alloy connected in parallel each having a cross-section of $25 \times 2\ mm^2$. The normal voltage drop is 75 mV. The alloy used having the following data : resistivity = $0.4 \times 10^{-6}\ \Omega\ m$, specific heat 500 J/kg- °C, specific gravity = 800 kg/m^3, rate of heat dissipation = 100 W/m^2·°C. Determine the maximum temperature rise at the time taken to reach 99 per cent of maximum value.*

Solution : Suppose l is the length of each strip expressed in m.

Area of each trip $a = 2 \times 25$ mm$^2 = 50 \times 10^{-6}$ m^2

Resistance of each strip $= \dfrac{\rho l}{a} = \dfrac{0.4 \times 10^{-6} \times l}{50 \times 10^{-6}} = 0.008\ l$

There are 8 strips in parallel

$\therefore$ Resistance of shunt $R_{sh} = \dfrac{0.008\ l}{8} = 0.001\ l$

Voltage drop across shunt $= I_{sh} \times R_{sh} = 1000 \times 0.001\ l$

$\qquad = 75 \times 10^{-3}$ (given)

$\therefore$ Length of each strip

$\qquad l = 0.075$ m $= 75$ mm

Volume of 8 strips $= 8 \times 75 \times 50 \times 10^{-6} = 0.03 \times 10^{-3}$ m^2

$\therefore$ Weight of strips $G = 8000 \times 0.03 \times 10^{-3} = 0.24$ kg

Total heat dissipating surface

$\qquad S = 8 \times 2\ (2 + 25) \times 75 \times 10^{-6} = 0.0324$ m^2

$\therefore$ Time constant $T_h = \dfrac{Gh}{S\lambda} = \dfrac{0.24 \times 500}{0.0324 \times 100} = 37$ s.

Loss in strips $Q = 1000 \times 75 \times 10^{-3} = 75$ W

Maximum steady temperature rise $\theta_m = \dfrac{Q}{S\lambda} = \dfrac{57}{0.0324 \times 100}$

$\qquad = 23.15°C.$

Suppose t is the time required to reach 99 per cent of the final steady temperature. From Eq. 4.51,

$\qquad \theta = \theta_m\ (1 - e^{-t/T_h})$

or $\qquad \theta/\theta_m = 0.99 = (1 - e^{-t/T_t})$ or t = 170.4 s.

Example 4.8 : *The heat dissipating surface of a 7.5 kW totally enclosed induction motor can be approximated as a cylinder of 0.6 m in diameter and 0.9 m in length. The motor can be considered to be made up of a homogeneous material weighting 375 kg and having a specific heat of 725 J/kg- °C. The specific heat dissipation from its surface is 12 W/m^2- °C.*

Find (i) the temperature rise of the machine at full load if the efficiency is 90 per cent. Also find the thermal time constant of the machine and (ii) the rotating of the machine for the same temperature rise if the machine were screen protected with a specific heat dissipation of 25 W/m^2- °C. Assume the same value for efficiency.

Solution : Total heat dissipating surface

$\qquad\qquad S =$ outer cylindrical surface + 2 $\times$ end

$\qquad\qquad\qquad$ surface

$\qquad\qquad\quad = \pi \times 0.6 \times 0.9 + 2 \times \pi/4 \times 0.6^2$

$\qquad\qquad\quad = 2.26$ m^2

$\qquad$ Efficiency $= \dfrac{\text{output}}{\text{output + losses}}$

or $\qquad 0.9 = \dfrac{7500}{7500 + \text{losses}}$

$\therefore$ Losses Q = 833 W,

From Eq. 4.26, final study temperature rise,

$\qquad\qquad \theta_m = \dfrac{Q}{S\lambda} = \dfrac{833}{2.26 \times 12} = 30.85°C$

From Eq. 4.29,

Time constant

$\qquad\qquad T_h = \dfrac{Gh}{S\lambda} = \dfrac{375 \times 725}{2.26 \times 12} = 10025$

$\qquad\qquad s = 2.79$ hours.

(ii)   We have, for screen protected machine,

$\qquad\qquad \theta_m = 30.85°C,\ S = 2.26$ m^2, $\lambda = 25$ W/m^2-°C

$\therefore$ From Eq. 4.26, allowable losses

$\qquad\qquad Q = \theta_m S\lambda = 30.85 \times 2.26 \times 25 = 1743$ W.

$\therefore$ Efficiency $= 0.9 = \dfrac{\text{output}}{\text{output + 1743}}$

or   output $= 15687$ W $= 15.7$ kW

Example 4.9 : *The temperature rise if a transformer is 25 °C after one hour and 37.5 °C after two hours of starting from cold conditions. Calculate its final steady temperature rise and the heating time constant. If its temperature falls from the final steady value to 40 °C in 2.5 hour when disconnected, calculate its cooling time constant. The ambient temperature is 30 °C.*

Solution :

When Heating : Since the transformer starts from cold conditions therefore its temperature rise is given by Eq. 4.29,

$\qquad\qquad \theta = \theta m(1 - e^{t/T_h})$

Now we have $\theta = 25°C$ at t = 1 hour

and $\qquad \theta = 37.5°C$ at t = 2 hours.

Putting these value in Eq. 4.29,

$\qquad\qquad 25 = \theta m\ (1 - e^{-t/T_h})$ ... (i)

$\qquad\qquad 37.5 = \theta m\ (1 - e^{-2/T_h})$ ... (ii)

From (i) and (ii)

$\qquad \dfrac{(1 - e^{-2/T_h})}{(1 - e^{-1/T_h})} = \dfrac{37.5}{25} = 1.5$

or $1 + e^{-1/T_h} = 1.5$ or $e^{-1/T_h} = 0.5$

or   Heat time constant $T_h = 1.44$ hour.

From (i), $25 = \theta_m\ (1 - e^{-1/T_h}) = \theta_m(1 - 0.5)$

$\therefore$ Final steady temperature rise $\theta_m = 25/0.5 = 50°C.$

When Cooling : Temperature rise after 2.5 hour, θ = 40 − 30 = 10°C. Since the transformer is disconnected its final steady temperature rise when cooling θ_n = 0.

Initial temperature rise is $\theta_i = \theta_m$ = 50°C

From Eq. 4.56, we have, $\theta = \theta_i\, e^{-t_e/T_e}$ or 10 = 50 $e^{-2.5/T_c}$

or cooling time constant T_c = 1.55 hour

Example 4.10 : *The initial temperature of a machine is 40°C. Calculate the temperature of the machine after 1 hour if its final steady temperature rise is 80°C and the heating time constant is 2 hours. The ambient temperature is 30°C.*

Solution : Initial temperature rise θ_i = 40 − 30 = 10°C

From Eq. 4.31, $\theta = \theta_m(1 - e^{-t/T_h}) + \theta_i\, e^{-t/T_h}$

$\therefore$　temperature rise after one hour

$$\theta = 80(1 - e^{-1/2}) + 10e^{-1/2} = 37.54\ °C$$

$\therefore$　Temperature of machine after one hour = 37.54 + 30

$$= 67.54°C$$

4.4 RATING OF MACHINES

- The rating of machines refers to the whole of the numerical values of electrical and mechanical quantities with their duration and sequences as signed to the machines by the manufactures and stated on the rating plate, the machine complying with the specified conditions. The duration of the sequence may be indicated by the qualifying term.

- The assignment of the rating is to be made by the manufacturer top specify the capabilities of the machine. Irrespective of whether the machine assigned to it by the machine manufacturer. In addition, since electrical machines have a time rate of temperature-rise and since the rise of temperature is limited according to standards, a qualifying term may be used with the term rating to give the duration fir which the machine may be run at the assigned values while complying with the standards.

- Where a machine is manufactured for general purposes, it is capable of supplying its power rating indefinitely and the qualifying term signifies this. Where a machine is manufactures with the intention that it may be used to supply varying loads or loads including periods of no-load or where the machine may be in a state of rest and de-energized, the qualifying term will signify this.

4.5 SELECTION OF MOTOR POWER RATINGS

- The selection of power ratings of motors for electric drives is a matter of economic and operational interest in any industry. This is because the proper selection of motor has a predominant effect on both capital and running costs of electric drives. A motor of insufficient capacity does not operate the drive satisfactorily on account of low output and efficiency. Since the motor is overloaded, it has a shorter life span and also a possibility of burn out on account of excessive temperature rise.

- On the other hand, if a motor of higher power rating is used, the motor is under utilized and therefore the economic efficiency of the installation is reduced and the drive becomes expensive and has large energy losses. Over motoring (using a motor of higher rating than is required by load) leads to higher capital costs and increased losses because of lower efficiency at reduced load. In a.c. drives, motors working at reduced loads to poor power factor leading to uneconomic loading of supply circuits and apparatus.

- In order to select the motor power rating properly, it is not only necessary to know the load under steady state conditions but also the loads that are met with transient conditions. For this purpose use is made of Load Diagrams (Time sequence graphs) which show the variation of motor torque, power and load current as function of time.

- The motors selected according to a given load diagram, should be fully loaded and in no case their temperature rise exceed the maximum permissible as specified by the manufacturer. In addition, the motors selected should operate properly during periods of overload and possess sufficient starting torque to accelerate the driven load up to the desired speed within the required time.

- In most of the situations, motors are selected on the basis of their heating since temperature rise of the motor is the prime factor in determining its life. The capacity of the motor is then checked for overload capacity. Motor power capacity is selected in accordance with the work the motor is desired to perform simultaneously ensuring that the motor will operate with permissible limits of mechanical loading.

- Since the motors operate under diverse operating (loading) conditions, it is pertinent here to define the Rating of a motor. The rating of a motor is the power output or the designated operating limit based upon certain definite conditions assigned to it by the manufacturer. An electrical motor is normally rated upon thermal basis of temperature rise i.e. maximum possible temperature at which the insulating materials may be operated without deterioration.

- In majority of industrial and other applications, motors carry constant load continuously. However, in may cases the motor operation is often stopped before the machine reaches its final steady temperature rise, or the load is reduced and the temperature is lowered on account or reduced losses.

4.6 COOLING OF ROTATING MACHINES

Bureau of Indian Standards (BIS 6362 – 1971) "Designation of Methods of Cooling of Rotating Electrical Machines" defines terms connected with cooling of Rotating Electrical Machines. Some of the commonly used terms are explained below :

- **Cooling :** A process by means of which heat resulting from losses occurring in a machine is given up to a primary coolant by increasing its temperature. The heated primary coolant by increasing its temperature. The heated primary coolant may be replaced by new coolant at lower temperature or may be cooled by a secondary coolant in some form of heat exchanger.

- **Primary Coolant :** A medium (liquid or gas) which, by being at a lower temperature than a part of a machine and in contact with it, removes heat from that part.

- **Secondary Coolant :** A medium (liquid or gas) which, being at a lower temperature than the primary coolant, removes the heat from the primary coolant in a heat exchanger.

- **Heat Exchanger :** A component intended to transfer heat from one coolant to another while keeping the two coolants separate (i.e. air cooled heat exchanger, water cooled heat exchanger, double wall, ribbed tubes, etc.).

- **Inner Cooled (Direct Cooled) Winding :** A winding which has either hollow conductors or tubes which form an integral part of the winding, through which the coolant flows.

- **Open Circuit Cooling :** A method of cooling in which the coolant is drawn from medium surrounding the machine, passes through the machine and then returns to the surrounding medium.

- **Closed Circuit Cooling System :** A method of cooling in which a primary coolant is circulated in a closed circuit through the machine, and if necessary through a

heat exchanger. Heat is transferred from the primary coolant to the secondary coolant either through the structural parts or in the heat exchanger.

- **Dependent Circulating Circuit Component :** A separate component in the coolant circulating circuit which is dependent for its operation on the operation of the main machine.

- **Independent Circulating Circuit Component :** A separate component in the coolant circulating circuit which is independent of the operation of the main machine.

- **Integral Circulating Circuit Component :** A component, in the coolant circulating circuit which forms part of the machine, and which can be replaced only by partially dismantling the main machine.

- **Machine Mounted Circulating Circuit Component :** A component in the coolant circulating circuit which is mounted on machine, and forms part of it, but which can be replaced without disturbing the main machine.

- **Separately Mounted Circulating Circuit Component:** A component in the coolant circulating circuit which is associated with a machine, but which is not mounted on or integral with it.

4.7 METHODS OF COOLING

- The factor which determines the size of a machine for a given duty is the temperature rise which occurs as a result of the various losses in the machine. Maximum allowable values of temperature rise in various parts of a machine have been standardized by International Electro-technical Commission. It is possible that, as new insulting materials suitable for withstanding higher temperatures are developed, these values will be revised upwards; but for the immediate future, the greatest gains of output from a given size of machine are likely to arise from improvements in the cooling techniques.

- Small electrical machines in the fractional horse power range may be cooled by natural means. In this method no external devices are used and the cooling is done by natural radiation and convection assisted by random air currents set up by rotor where the frames are open. But all modern electrical machines are characterized by large losses per unit area of surfaces of the machine, which dissipate heat into the ambient medium and hence artificial cooling methods are necessary in order to avoid excessive temperature rises during machine operation.

- In most cases the cooling of electrical machines is done by air streams and this cooling is called "Ventilation". In high speed machines such as turbo-alternators, hydrogen is used for cooling. There are machines in which water is used for cooling.

4.8 COOLING SYSTEM

According to BIS : 4722-1968 "Specification for Rotating Electrical Machines" the cooling systems are classified into three types depending upon the origin of cooling. :

1. **Cooling System :** The machine, is cooled by natural air currents set up either by rotating parts or due to temperature differences. The machine thus is cooled without the use of a fan by the movement of air and radiation.

2. **Natural Cooling :** The machine is cooled by cooling air driven by a fan mounted on the rotor or one driven by it.

3. **Separate Cooling :** The machine is cooled either by a fan not driven by its shaft, or it is cooled by a cooling medium other than air put into motion by means not belonging to the machine.

The cooling of machines, according to the manner of cooling is of following types :

- **Open Circuit Ventilation :** The heat is given by directly to the cooling air through the machine which is being continuously replaced.

- **Surface Ventilation :** The heat is given up by the cooling medium from the external surface of a totally enclosed machine.

- **Closed Circuit Ventilation :** The heat is transferred to the cooling medium through an intermediate cooling medium circulating in a closed circuit through the machine and a cooler.

- **Liquid Cooling :** Parts of the machine carry water or another kind of liquid flowing through them, or they are immersed into a liquid.

- **Inner Cooling of Windings :** This is of two types :

    (i) Inner Gas Cooling : One or all windings are cooled by a gas, for instance hydrogen, flowing internally through the conductors or coil.

    (ii) Inner Liquid Cooling : One or all windings are cooled by a liquid, for instance, water flowing internally through the conductors or coils.

4.9 ENCLOSURES FOR ROTATING ELECTRICAL MACHINES

The problem of ventilation in rotating electrical machines is closely lined with the types of enclosures used. The various types of enclosures used are :

- **Open Machine :** One in which there is no restriction to ventilation other than that necessitated by good mechanical construction.

- **Open Pedestal Machine (OP) :** An open machine which has pedestal bearing supported independently of the machine frame.

- **Open End-Bracket Machine (OEB) :** An open machine having end-brackets of which the bearings from an integral part.

- **Protected Machine (P) :** A machine in which the internal rotating parts and live parts are protected mechanically from accidental or inadvertent contact, while ventilation is not materially impeded.

- **Screen Protected Machine (SP) :** A protected machine in which the ventilating openings are not less than 64.5 mm^2 in area. Such protection may be provided by screens of wire mesh, expanded metal, perforated metal or other suitable covers. The use if openings are liable to become closed in service.

- **Drip-Proof Machine (DP) :** A protected machine in which the openings for ventilation are so protected as to exclude vertically falling water or dirt.

- **Splash-Proof Machine (SPLP) :** A protected machine in which the ventilating openings are so constructed that drops of liquid or solid particles falling on or reaching any part of them machine at any angle between the vertically downward direction and 100° from that direction cannot enter the machine, whether the machine is running or at rest, by splashing, or otherwise, either directly or by striking and running along a surface.

- **Hose-Proof Machine (HSP) :** A protected machine so enclosed as to exclude water whether the machine is running or at rest, when washed by a hose having a 9.5 mm diameter nozzle with a maximum pressure of 3.5 kg/cm^2 for a period not exceeding 30 seconds, from a minimum distance of 1.8 meters.

- **Pipe-Ventilated or Duct-Ventilated Machine :** A machine in which there is a continuous supply or fresh ventilating air, the frame being so arranged that the ventilating air may be conveyed to and/or from the machine through pipes or ducts attached to the enclosing case :

> A pipe-ventilated or duct-ventilated machine may be one of the following three types :

   (i) With provision for inlet duct, only.

   (ii) With provision for inlet and outlet ducts.

   (iii) With provision for outlet duct only.

> A pipe-ventilated or duct-ventilated machine may be cooled by one of the following means :

   (i) Self-ventilation (PV).

   (ii) Forced-draught with air supplied by external pressure (PVFD).

   (iii) Induced draught with air drawn through the machine by external means (PVID).

- **Totally Enclosed Machine (TE) :** A machine so constructed that the enclosed air has no connection with the external air but is not necessarily 'air-tight'.

- **Totally Enclosed Fan-Cooled Machine (TEFC) :** A totally enclosed machine with cooling augmented by a fan, driven by the motor itself, blowing external air over the cooling surface and/or through the cooling passages, if any, incorporated in the machine.

- **Totally Enclosed Separately Air-cooled Machine (TESAC) :** A totally-enclosed machine with cooling augmented by a separately-driven fan blowing air over the cooling surface and/or through the cooling passages, if any, incorporated in the machine.

- **Totally Enclosed Water or other Liquid-Cooled Machine (TEWC) :** A totally enclosed machine with cooling augmented by water-cooled or other liquid-cooled surfaces embodied in the machine itself.

- **A Totally Enclosed Closed Air Circuit Machine :** A totally enclosed machine having special provision for cooling the enclosed air by passing it through its own cooler, usually external to the machine. The cooler may be of any recognised form using :

> Air (CACA)   > Water (CACW)

> Other suitable cooling medium.

- **Totally Enclosed Closed Gas Circuit Machine (CGGW) :** A totally enclosed machine cooled by gas other than air, the cooling has being circulated through associated water-cooled gas coolers.

- **Weather-Proof Machine (WP) :** A machine so constructed that it can work without further protection from weather conditions specified by the purchaser.

- **Watertight Machine (WT) :** A machine so constructed that it will withstand, without damage or sign of leakage, complete immersion in water to a depth of not less than 1 m, or subjection to an external water pressure of 0.1 kg/cm^2 for a period of one hour. The test for watertightness shall be made with the machine stationary and the temperature of the machine shall not exceed the temperature of the water in which it is immersed.

- **Submersible Machine :** A machine capable of working for an indefinitely long period when submerged under a specified head of water.

- **Flame Proof Machine (FLP) :** A machine which compiles with the requirements of IS : 2148 – 1962 "Specification for flame-proof enclosures of electrical apparatus".

4.10 METHODS OF VENTILATION OF COOLING

- It has already been stated that energy conversion in electrical machines is associated with the production of losses. The losses generate heat and temperature rise inside the machine. The temperature rise if allowed to increase beyond a permissible limit, depending upon the type of insulation used, will deteriorate the insulation and renders the machine useless. Thus it is necessary to provide suitable ventilation and cooling for the machine. So that the temperature rise at any part of the machine does not go beyond the permissible limit.

- The rating of an electrical machine consequently depends upon how fast the heat is transferred from inside the machine and dissipated to the ambient medium. For increasing rating of an electrical machine wither insulating material of higher temperature limit is developed and used or sophisticated cooling techniques are employed.

- Very small electrical machines are cooled by natural means with no external device like fans etc. As the size of the machine increases the losses and consequently temperature rise increases faster than the increase in the surface area thus requiring artificial cooling methods.

The cooling of electrical machines by means or an air stream is called ventilation of the machines.

The cooling systems can be grouped into three types :

1. Natural Cooling :

The machine is cooled by air movements set up in the machine due to its rotation or due to the temperature difference between inside parts and the ambient air.

2. Self Cooling :

The machine is cooled by air, blown by fan integrally built with the rotor or mounted on the shaft.

3. Separate Cooling

The machine is cooled by air, blown by fan driven by separate machine.

Further, the ventilation of machine can be classified in to three types according to the schemes of ventilation incorporated in the machine.

1. Open Circuit Ventilation :

Here the heat is given up directly to the cooling air, the air is being replaced continuously.

The open circuit ventilation can be divided in to two types in accordance with how the air is brought in to the machine.

(i) Induced Ventilation : In this scheme a fan produces a decreased pressure of air inside the machine and air from outside is sucked into the machine owing to the atmospheric pressure. The air is circulated through the machine and pushed out by the fan into the atmosphere. The fan is mist normally arranged inside the machine as shown in Fig. 4.4 The scheme is called induced ventilation with internal fan. The fan can be arranged externally also, to be called as induced ventilation with external fan.

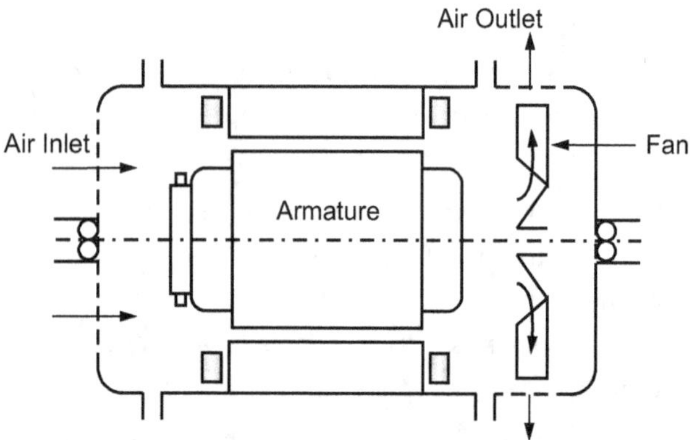

Fig. 4.4 : Induced ventilation with internal fan arranged in a d.c. machine

(ii) Forced Ventilation : Here a fan sucks the air from outside and forces it into the machine and finally pushed out into the atmosphere. Again the fan is mostly arranged inside the machine as shown in Fig. 4.5 The scheme is called forced ventilation with internal fan. The fan can be arranged externally also, to be called as forced ventilation with external fan.

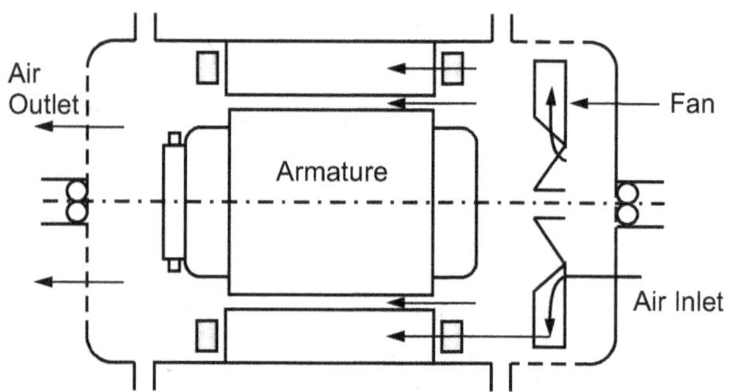

Fig. 4.5 : Forced ventilation with external fan arranged in a d.c. machine

2. Closed Circuit Ventilation : In closed circuit ventilation system heat is transferred through an intermediate ventilating medium normally air or hydrogen circulating in a closed circuit between the interior of the machine and a cooler.

3. Surface Ventilation : The heat is transferred from inside of the machine, by cooling medium, to the external surface of a totally enclosed machine. The external surface is being cooled by natural means or mainly by air blown by fan.

4.11 COOLING-AIR CIRCUIT

The ventilating systems can be further classified in to four types in accordance with the provision of cooling ducts and how the air passes over the heated parts of the machine.

1. Radial,   2. Axial,
3. Combined Radial and Axial,   4. Multiple inlet

4.11.1 Radial Ventilating System

- This is commonly employed because the movement of rotor induces natural centrifugal motion of the air. The movement of the air can be increased, if required, by rotor fans. Fig. 4.6 shows radial system for small machines say below 20 kw rating. The end brackets are so arranged and shaped to guide air over the overhang and the back of the core. For large size machines having longer core lengths,. subdivision of the core is necessary, as shown in Fig. 4.7, and the air paths through the radial ducts are in parallel with those across the over hang. A high rate of heat dissipation is possible.

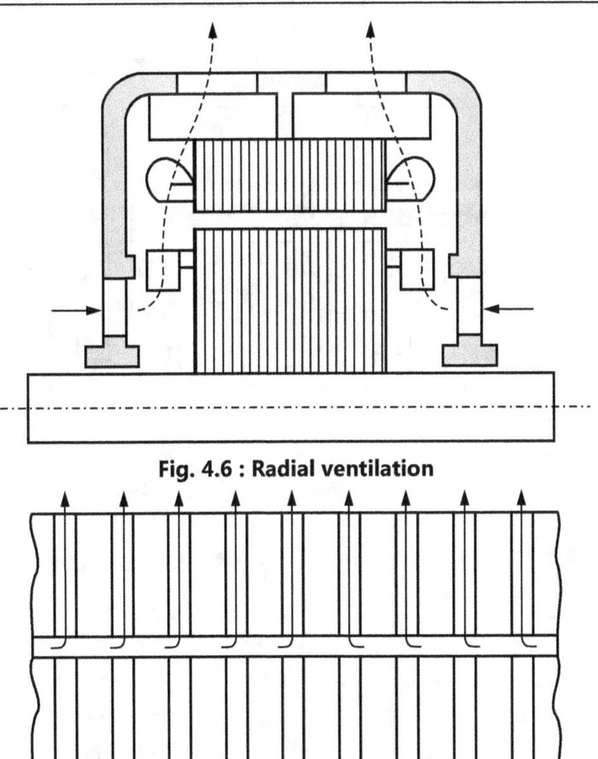

Fig. 4.6 : Radial ventilation

Fig. 4.7 : radial ventilation ducts in long cores

4.11.2 Axial Ventilating System

- This system is suitable output and high speed machines. This is because in high speed machines, a solid rotor construction with restricted spider is used in order to avoid centrifugal stresses and this restricts the provision of radial ventilating ducts and hence axial ventilation has to be used Fig. 4.8 shows a method of applying axial ventilation to a small machine with plain cores. To increase the cooling surface, holes may be punched in the core plates to form through-ducts where considerable heat dissipation occurs. This greatly improves the cooling, but requires a large core diameter for the increased core depth necessary.

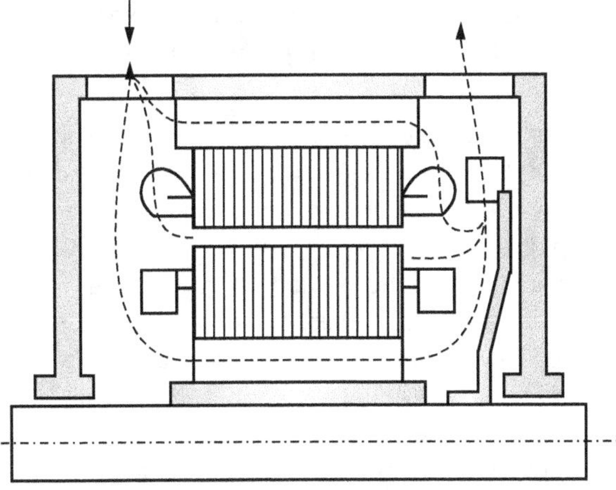

Fig. 4.8 : Axial ventilation

4.11.3 Combined Radial and Axial Ventilation System

- This method is normally used for large motors and small turbo-alternators. Fig. 4.9 and guided to pass through the ducts by baffling the fan end of the rotor spider. The shaft-mounted fan forces out the heated air.

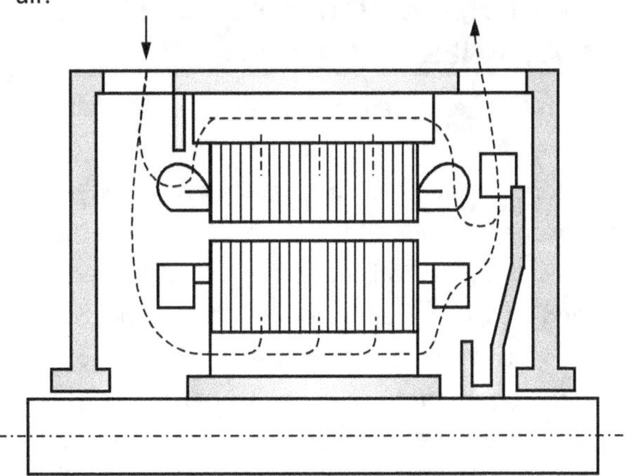

Fig. 4.9 : Combined Radial & Axial Ventilating System

4.11.4 Multiple-Inlet Ventilation System

- The previous method of axial ventilation can not be used for turbo-machine having long core lengths, because by the time the cooling air reaches the central parts of the machine it has already become hot. Thus in large machines multiple inlet system is adopted.

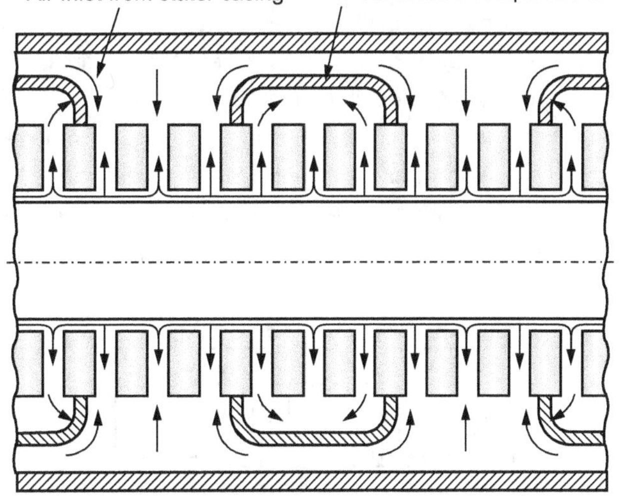

Fig. 4.10 : Multiple inlet system

- In this system the stator frame and core is divided in to a number of compartments. These compartments are used alternately as inlet and outlet chambers. In the inlet chambers the air is forced radially inwards where as in the outlet chambers the air is forced outwards. In this system it is taken case that each part of the machine receives cool air. Air under pressure is forced in to the stator casing from where it enters the inlet chambers for on ward passing through the ducts. The heated air is drawn out from the outlet chambers and is subjected to the coolers where it is cooled for recirculation.

4.12 COOLING OF TOTALLY ENCLOSED MACHINES

- In a totally enclosed machine the inside air has no connection with the outside. Therefore the heat developed inside the machine is dissipated in to the atmosphere through the external surface of the frame, i.e. by surface ventilation. The outer heated surface having ribs is normally not left to cooling by natural means otherwise the heat dissipation is poor leading to low rated machine. The heated outer surface is cooled buy blast of air by a shaft mounted fan. Now two types of fan cooling are adopted :

  - As shown in Fig. 4.11, shaft mounted fans external to the working parts of the machine blows air over the across through a space between the main housing and a thin cover plate. Internal air circulation is produced by an internal fan, This avoids the temperature gradient across the air gap.

  - Another arrangement is shown in Fig. 4.12, an internal fan circulates the heat to the carcass. Air is also blown over the outside of the carcass to improve the dissipation. This outer fan is enclosed by another cover in order to secure required direction of air flow. This type of ventilation is the most common one and is adopted for machines up to 25 Kw.

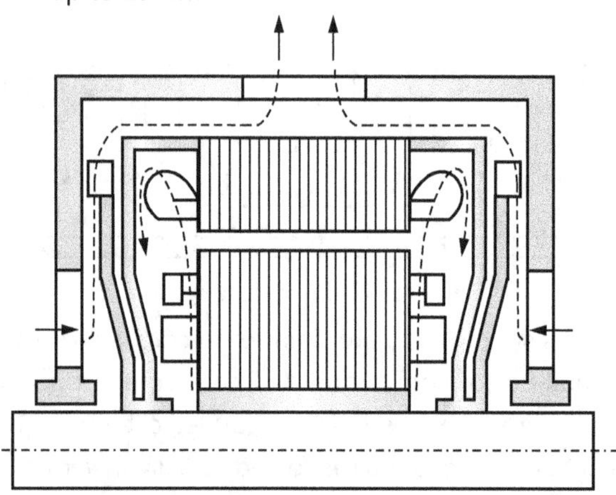

Fig. 4.11 : Ventilation of totally enclosed machine with external fan

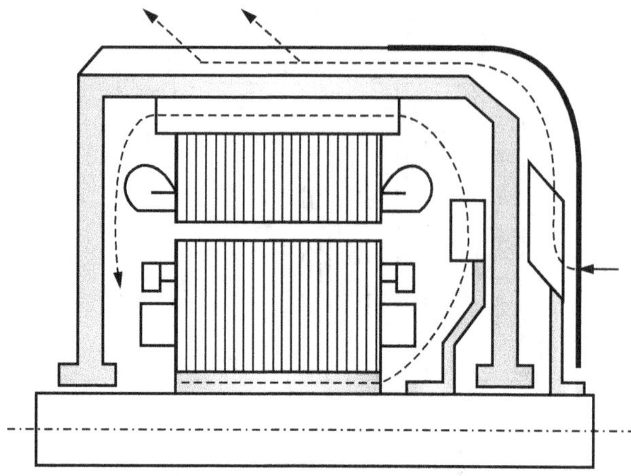

Fig. 4.12 : Ventilation of totally enclosed machine with internal fan

4.12.1 Cooling of Large Sized Machines (Turbo-Alternators)

- Among the large sized machines, cooling of turbo-machines has always posed most complex problems. Turbo-machine coupled to a steam turbine is basically a high speed machine having limited diameter but long core length. If open circuit ventilation is adopted, large volumes of air is required, it is necessary to clean the air with suitable filters to prevent clogging of ducts with dust.

- The air brought in must be moisture free. Thus for requirements of several tons of air per hour, depending upon the rating of the machine, the operating cost will be very heavy. Further it can not ensure dust free entry of air.

4.12.2 Closed Circuit Air Cooling

- A complete means of securing clean cooling air is by closed circuit ventilation system. Here the same volume of air passes through a closed circuit again and again. This circuit consists of various ducts in the machine, together with a chamber below containing the fans, air coolers, drying agents etc.

- The cold air after extracting the heart from the machine becomes hot and is passed through a water cooled heat exchanger and then returned to the machine by a centrifugal fan. Closed circuit air cooling is applicable only up to a maximum of say 50 MW. For machines of higher rating large volumes of air is required and is to be circulated requiring hundreds of kw of fan power. Because of this alone there is no scope of increasing the rating to cope up with the increased losses. Therefore an alternative cooling medium had to be searched. The search was ended with hydrogen gas as a cooling medium.

4.12.3 Closed Circuit Hydrogen Cooling

- There are certain obvious advantages when hydrogen is used as a coolant in place of air under closed circuit cooling :

Advantages

- **Reduced Fan Power and Reduced Windage Loss and Noise :** The density of hydrogen is 7% ($\cong 1/14$) of that of air; obviously requiring mush less fan power circulation ; reduced windage loss and less noise. There is an improvement in efficiency by 0.5 to 1%.

- **Better Cooling Medium :** Specific heat of hydrogen is 14 times that of air; thermal conductivity 7 times and heat transfer 1.5 times. So more rapidly extracts and gives way heat, reduced temperature gradients. Therefore an increased output can be taken from a given frame size. The increase in rating is by about 15% at a pressure of 1 atm, by about 30% at 2 atm and 40% at 3 atm.

- **Increased Life :** Thermal conductivity of hydrogen is higher that of air which is of the same order as that of winding insulations. Therefore if there are pockets filled with hydrogen, heat conduction through them is as good as through insulation and thus local high temperatures are avoided. Such pockets under air cooling creates hot spots and insulation breakdown. This problem as has been seen is avoided under hydrogen and consequently increase in the life of insulation.

  Insulation corona which is present when air is used as a coolant in high voltage machines is greatly reduced under hydrogen, thus again increasing the life of insulation.

- **Elimination of Fire Hazed :** Fire hazards are eliminated as hydrogen does not support combustion so long as the hydrogen/ air mixture exceeds 3/1.

- **Smaller Size of Heat-Exchanger :** Smaller size of heat-exchanger /cooler is required to cool the heated hydrogen.

- **An Important Precaution :** Hydrogen if mixed with air forms an explosive mixture over a wide range. (Hydrogen to air; 4% to 76%). Therefore the machine frame must be gas-tight and explosion proof.

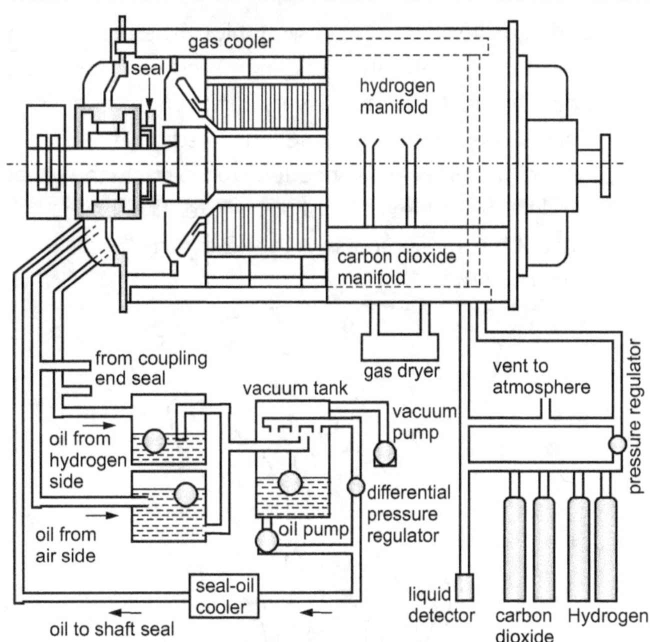

Fig. 4.13 : Hydrogen cooling circuit

- Oil-film gas seals are used to ensure gas tight frame. Internal explosion is to be avoided by

  - ➤ Maintaining gas pressure above atmospheric by an automatic regulating and reducing valve controlling the supply from normal gas cylinders. Thus any leakage if occurs us outwards where hydrogen is quickly dissipated to the atmosphere. The hydrogen gas pressure is generally 1 to 2 kg per cm^2.

  - ➤ While filling or emptying the casing of the machine an explosive hydrogen air mixture is avoided by first displacing the air by carbon-dioxide before admitting hydrogen; and the process is reversed for emptying (i.e. carbon-dioxide is used to replace hydrogen before emptying the casing).

- Fans mounted on the rotor shaft circulate hydrogen through the ventilating ducts and gas coolers. Hydrogen cooling results in substantial increase in rating for a given temperature rise, and along with the reduced windage loss may add 0.5 to 1% to the efficiency of a say 100 to 120 Mw generator. Fig. 4.13 shows the machine along with the auxiliary equipment.

4.12.4 Direct Cooling

- We have seen that a large size turbo-generator machine is provided with both axial and radial ventilating ducts (see Fig. 4.14) with multiple inlet closed circuit hydrogen cooling; hydrogen pressure maintained at 2-3 atm. pressure. But for machines of 100 Mw or more the temperature gradient over the conductor insulation is still very high. Therefore direct contact between the cooling medium and the conductor material is still very high. Therefore direct contact between the cooling medium and the conductor material is needed to extract the heat quickly. This type of cooling where the conductor material has direct access with the coolant is called as direct cooling.

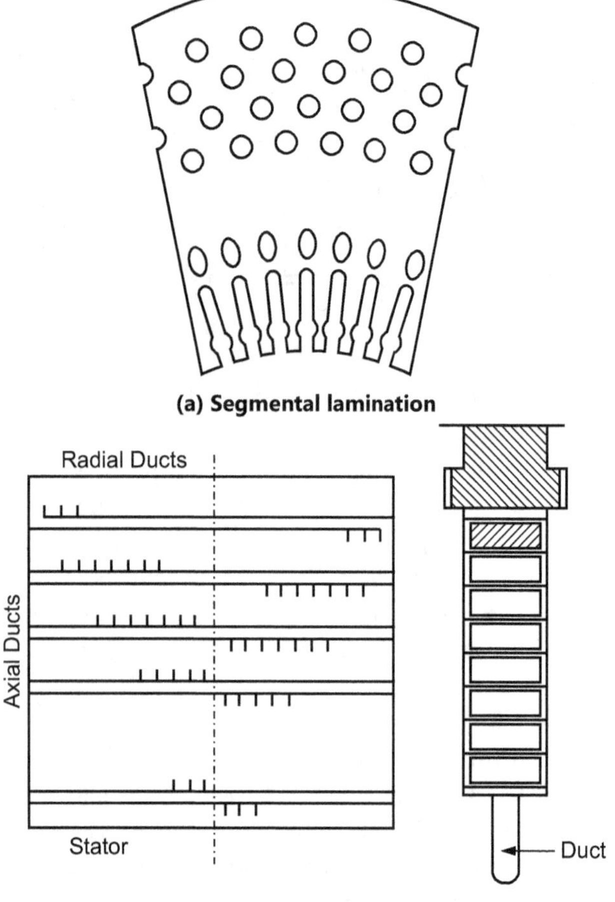

(a) Segmental lamination

(b) Showing radial and axial ducts   **(c) Showing rotor slot with duct**

Fig. 4.14 : Large size-turbo-machine cooling

Advantages :

* Due to quick transfer of heat, rating of the machine can be considerably increased.

* Since coolant comes in direct contact with the conductor, the temperature gradient across slot insulation, teeth and surface barrier are considerably reduced.

* The coolant used normally are hydrogen and water.

    (i) Direct Gas (Hydrogen) Cooling : Here both stator ad rotor conductors are made hollow and hydrogen is forced through them from one end to the other. Fig. 4.15 shows a rotor slot of a direct hydrogen cooled turbo-machine. The conductors comprise rectangular ducts/tubes receiving hydrogen from a cooling circuit other than that of stator. The hydrogen is admitted to the tubes through insulating flexible connections at the ends from a pump which is mounted on the out board end of the shaft. The slot tubes are electrically connected at the overhang by suitably shaped copper bars forming inlet or outlet ports. The hollow conductors are of hard drawn silver bearing copper with synthetic resin bonded glass-cloth laminate insulation.

With direct hydrogen cooling of stator and rotor windings and hydrogen cooled stator cores, the machine permits much higher electric loading say of the order of 300 to 500 Mw. With higher rating of the order if 1000 Mw even this type of cooling is not enough to cope up with heavy flow rate and pressure of hydrogen to extract the rotor winding loss of the order of say 5 Mw. Therefore direct water cooling is employed for such case.

(ii) Direct Water Cooling : Turbo-generators of the highest possible ratings thought of are likely to have hydrogen cooled stator cores and direct water cooling of stator and rotor windings. Water as a coolant has superior heat transfer capacity and the viscosity of pure water is very small requiring not high pressure heads to circulate within small tubes.

One of the main problem of direct cooling is to device flexible water-tube connections with insulation against high voltages of the windings. Another problem is to maintain low conductivity of the coolant water. Direct water cooling of the rotor conductors poses still greater problems because of the obvious reasons. Fig. 4.16 shows rotor slot with conductor arrangement. The rotor slot is narrow at the bottom side to allow for greater tooth width so as to withstand large centrifugal forces. The water pumping pressure is so maintained that the velocity of water hardly exceeds 1.5 m/sec. The maximum limit is 2.5 m/sec. The speed is kept low in order to avoid erosion and cavitation. Reservoir tanks maintain the necessary water levels. The purity of water is constantly monitored and can be controlled by a demineralization unit.

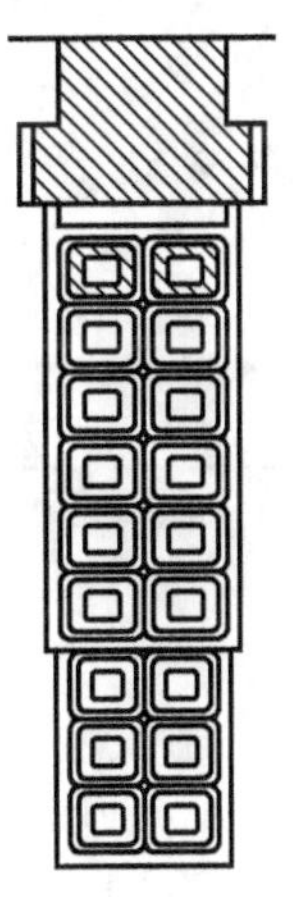

Fig. 4.15 : Slot of a direct hydrogen cooled turbo-rotor

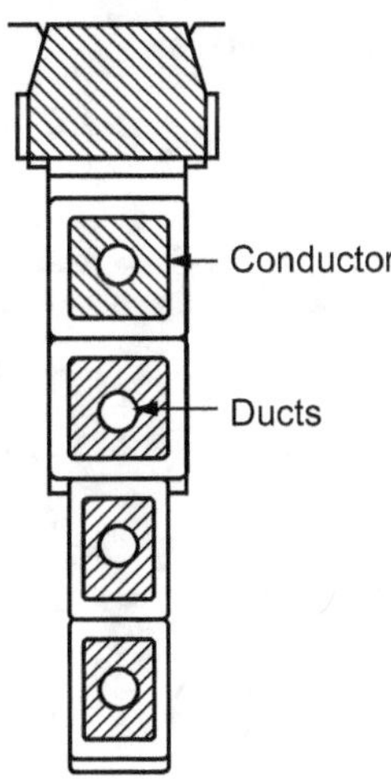

Fig. 4.16 : Slot of a direct water cooled turbo-rotor

4.13 QUANTITY OF COOLING MEDIUM (COOLANT)

The quantity of cooling medium required to absorb losses of machines can be calculated with the help of expressions derived below :

Let $\quad Q$ = loss to be dissipated, kW;

$\quad\theta_i$ = inlet temperature rise of cooling medium, °C;

$\quad\theta$ = temperature rise of cooling medium, °C;

$\quad H$ = barometric height, mm of mercury ;

$\quad P$ = pressure, N/m^2

4.13.1 Air

Heat to be carried away

$$= Q \text{ kW} = Q \times 10^3 \text{ W}$$

$$= W \times 10^3 \text{ J/s}$$

Let $\quad c_p$ = specific heat of air at constant pressure, J/kg-°C

We have, heat = weight × specific heat × temperature rise.

Heat per second = weight of air in kg/s × specific heat in j/kg –°C × temperature rise in °C.

$$\therefore \text{ Weight of air } = \frac{Q \times 10^3}{c_p\theta} \text{ kg/s.}$$

Let volume of 1 kg air at N.T.P. (0°C i.e. 273 K, and 760 mm of mercury) be V m^3.

$$\therefore \quad \text{Volume of air required at N.T.P.} = \frac{Q \times 10^3}{c_p\theta} V \text{ m}^3\text{/s}$$

At actual working conditions the temperature of air is θ_i°C or $(\theta_i + 273)$ K and the pressure is H mm of mercury. Applying Kelvin's law, we have :

$$\frac{P_1 V_1}{T_1} = \frac{P_2 V_2}{T_2}$$

Volume of air under actual working conditions is therefore:

$$V_a = \frac{Q}{c_p\theta} V \times \frac{\theta_i + 273}{273} \times \frac{760}{H} \times 10^3 \text{ m}^3\text{/s} \quad (4.38)$$

The values for day air are :

$$c_p = 0.2375 \text{ cal/gm-°C} = 995 \text{ J/kg-°C}$$

and $\quad V = 0.775$ m^3.

Substituting these values in Eq.

$$V_a = \frac{Q}{995\ \theta} \times 0.775 \times \frac{\theta_i + 273}{273} \times \frac{760}{H} \times 10^3 \text{ m}^3\text{/s}$$

$$= 0.78 \frac{Q}{\theta} \times \frac{\theta_i + 273}{273} \times \frac{760}{H} \text{ m}^3\text{/s}$$

Power required for blowing air at a rate V_a m^3/s at a pressure of P N/m^2, = PV_a watt.

$\therefore$ Power required by fan blowing air

$$P_f = P \frac{V_a}{\eta_f} \times 10^{-3} \text{ kW} \qquad \dots (4.39)$$

where $\quad \eta_f$ = fan efficiency.

Typical values for a.c. generators are :

$\quad P$ = 1000 to 2000 N/m^2 or higher for closed circuit ventilation

$\quad \theta$ = 20°C and θi = 25°C

$\quad \eta_f$ = 0.2 to 0.4.

4.13.2 Hydrogen

Specific heat of hydrogen at constant pressure = 3.4 cal/gm°C and 1 kg of dry gas occupies a volume of 11.3 m^3 at 273 K and 760 mm of mercury.

Processing as in the case of air :

Volume of hydrogen required for Q kW loss with a temperature rise θ°C,

$$V_H = 0.8 \frac{Q}{\theta} \cdot \frac{(\theta_i + 273)}{273} \cdot \frac{760}{H} \text{ m}^3\text{/s} \qquad \dots (4.40)$$

H for hydrogen is 2000 – 2500 mm.

4.13.3 Water

We have,

$$Q \text{ kW} = 1000 \text{ Q J/s.}$$

The specific heat of water is 4.18 × 10^3 J/kg-°C and so the volume of water required per second for absorbing Q kW loss with a temperature rise of θ°C,

$$V_\omega = \frac{1000\ Q}{4.18 \times 1000} \times \theta \text{ kg/s}$$

$$= \frac{0.24\ Q}{\theta} \text{ } l\text{/s} \qquad \dots (4.41)$$

4.13.4 Oil

The amount of oil required per second for a dissipation of Q kW loss with a temperature rise of $\theta°C$,

$$V_0 = \frac{0.24\ Q}{(0.35\ \text{to}\ 0.5)\ \theta}\ l/s. \qquad \qquad ...(4.42)$$

Example 4.11 : *A 50 MVA turbo-alternator has a total loss of 1500 kW. Calculate the volume of air required per second and also the fan power if the temperature rise in the machine is to be limited to 30 ℃. The other data gives is :*

    Inlet temperature of air = 25°C,

    Barometric height = 760 mm of mercury,

      Pressure = $2\ kN/m^2$,

    Fan efficiency = 0.4.

Solution : Given :

$$Q = 1500\ kW,\ \theta i = 25°C,\ \theta = 30°C,$$
$$H = 760\ mm.$$

In the absence of other data we assume that specific heat of air at constant pressure $c_p = 995\ J/kg\text{-}°C$ and volume of 1 kg of air at N.T.P. is $V = 0.775\ m^3$.

∴   Volume of air

$$V_a = 0.78 \times \frac{1500}{30} \times \frac{25+273}{273} \times \frac{760}{760}$$

$$= 42.6\ m^3/s.$$

$$\text{Fan power} = P_f = P\frac{V_a}{\eta_f}\ 10^{-3}$$

$$= \frac{2000 \times 42.5}{0.4} \times 10^{-3} = 212.5\ kW$$

Example 4.12 : *A turbo-alternator runs on test at a continuous rated load of 30 MVA with a power factor of 0.8. The following cooling air measurements are taken :*
Volume of cooling air measured at intake = 30 m³/s,
Intake air temperature = 15 ℃,
Outlet air temperature = 45 ℃,
Barometric reading = 750 mm of mercury.

(a) Find the efficiency of the machine, taking the specific heat of air at constant pressure as 1000 J/kg- ℃, and the volume of 1 kg of air at 0 ℃ and a pressure of 760 mm of mercury as 0.78 m³.

(b) Calculate the amount of cooling water in litre per second to cool the air, assuming the temperature rise of water to be 8 ℃.

Solution : Given :

(a) $\theta = 45 - 15 = 30°C$

        $\theta_i = 15°C,\ H = 750\ mm\ of\ mercury,$

        $V_a = 30\ m^3/s$

        $c_p = 1000\ J/kg\text{-}°C,\ V = 0.78\ m^3$

Volume of air

$$V_a = \frac{Q}{c_p\theta}V \times \frac{\theta_i+273}{273} \times \frac{760}{H} \times 10^3\ m^3/s$$

or $\text{loss}\ Q = V_a\frac{c_p\theta}{V} \times \frac{273}{\theta_i+273} \times \frac{H}{260} \times 10^{-3}\ kW$

$$= 30 \times \frac{1000 \times 30}{0.78} \times \frac{273}{15+273} \times \frac{750}{2760}$$

$$\times 10^{-3}$$

$$= 1080\ kW$$

Output $= 30 \times 10^3 \times 0.8 = 24 \times 10^3\ kW$

∴ Efficiency $= \dfrac{\text{output}}{\text{output + losses}} = \dfrac{24000}{24000 + 1080}$

$$= 95.7\%$$

(b) Losses $Q = 1080\ kW$

Temperature rise of water $\theta = 8°C.$

From eq. 4.41, amount of water,

$$V_\omega = \frac{0.24\ Q}{\theta} = \frac{0.24 \times 1080}{8} = 23.4\ l/s.$$

Example 4.13 : *The losses of a 60 MW hydrogen cooled alternator on full load amount to 750 kW. The flow of hydrogen from the coolers is 10 m/s at 2000 mm of mercury gauge pressure above atmosphere, which is 760 mm of mercury. The temperature of hydrogen leaving the coolers is 25 ℃. Determine the temperature rise of hydrogen assuming specific heat of hydrogen at constant pressure to be 12540 J/kg- ℃ and weight of 11.2 m³ of hydrogen at 0 ℃ and 760 mm of mercury to be 1 kg.*

Solution : Hydrogen enters the machine after leaving the coolers.

∴   Volume of hydrogen entering the machine

        = volume of hydrogen leaving the coolers

        $= 10\ m^3/s$

Eq. 4.38 has been derived for air bit in fact it is a general equation which is applicable to any gaseous cooling medium with relevant quantities substituted.

∴   Volume of hydrogen

$$V_H = \frac{QV}{c_p\theta} \times \frac{\theta_i+273}{273} \times \frac{760}{H} \times 10^3\ m^3/s$$

∴   Temperature rise $\theta = \dfrac{QV}{c_pV_H} \times \dfrac{\theta_i+273}{273} \times \dfrac{760}{H} \times 10^3$

$$= \frac{750 \times 11.2}{12540 \times 10} \times \frac{25+273}{273} \times \frac{760}{2760} \times 10^3$$

$$= 20°C$$

Example 4.14 : *The total losses in a 40 MVA transformer are 200 kW. The oil for cooling the transformer is circulated by a pump. The oil after taking up heat from the transformer tank goes to the radiators where it is cooled by circulation of water. Calculate the amount of oil which is raised by 20 ℃ in its passage through the tank. Also calculate the amount of water required for cooling oil in radiators. The rise in temperature of cooling water is 10 ℃. Assume that 20 percent of losses are dissipated by tank walls.*

Solution :

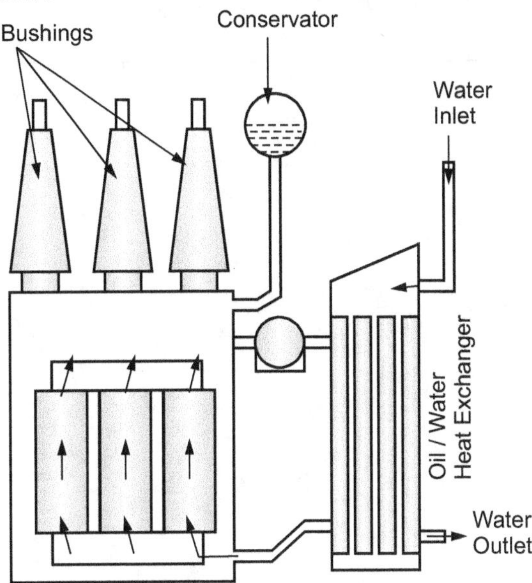

Fig. 4.17 : Oil forced water cooled transformer

Amount of oil :

Heat taken up by oil

$$= 0.8 \times 200$$
$$= 160 \text{ kW}$$

$\therefore$ $Q = 160$ kW and $\theta = 20°C$

From Eq. 4.43, oil required

$$V_o = \frac{0.24 \times 160}{0.4 \times 20}$$

$$= 4.8 \text{ } l/s \text{ (assuming } c_p = 0.4)$$

Amount of water : $Q = 160$ kW, $\theta = 10°C$

From Eq. 4.42, water required

$$V_\omega = \frac{0.24 \times 160}{10} = 3.84 \text{ } l/s$$

Example 4.15 : *A 15 MVA transformer has an iron loss of 80 kW and a copper loss of 120 kW at full load. The tank dimensions are 3.5 × 3.0 × 1.4 metre. The transformer oil is cooled by 3 litre of water per second passed through a cooling coil. Estimate the average temperature rise of the tank if the difference of temperature of water at the inlet and the outlet is 15 ℃. The specific loss dissipation from tank walls is 10 W/m² - ℃.*

Solution :

Total losses = 120 + 80 = 200 kW.

Out of this loss of 200 kW, some loss is dissipated by tank walls and the rest is taken up by circulating water.

Rise in temperature of water = 15°C

Amount of water per second V_ω = 3 litre.

From Eq. 4.65, amount of water $V_w = \dfrac{0.24 \text{ } Q}{\theta}$

Or heat taken away by water $Q = \dfrac{V_\omega \theta}{0.25} = \dfrac{3 \times 15}{0.24}$

$$= 187.5 \text{ kW}$$

The rate of the loss is to be dissipated by tank walls.

Loss dissipated by walls = 200 − 187.5 = 12.5 kW

Area of tank walls (neglecting top and bottom surfaces)

$$S = 2 \times 3.5(3 + 1.14)$$
$$= 30.8 \text{ m}^2$$

Temperature rise of tank

$$\theta = \frac{Q}{S\lambda} = \frac{12.5 \times 100}{30.8 \times 10} = 40.6 \text{ °C}$$

Example 4.16 : *A 500 MW direct water cooled turbo-alternator has a stator copper loss of 800 kW. The water inlet temperature is 38 ℃ and the outlet temperature is 68 ℃. Calculate amount of water required per second. Also calculate the area of water duct in each subconductor if there are 48 slots with 2 conductors per slot and each conductor is subdivided into 32 sub-conductors. The velocity is not to exceed 1.0 m/s.*

The pumping pressure is 300 kN/m², calculate the power of water if its efficiency is 0.6.

Solution : Temperature rise of water $\theta = 68 - 38 = 30°C$.

The volume of water required

$$V_\omega = \frac{0.24 \text{ } Q}{\theta} = \frac{0.24 \times 800}{30}$$

Total number of stator conductors = 2 × 48 = 96.

Total number of subconductors = 96 × 32.

Volume of water required for each sub-conductor

$$= \frac{6.4}{96 \times 32} = 0.00208 \text{ } l/s$$

EXERCISE

1. What are the sources of heat generation in a rotating electrical machine.

2. Why it is necessary to cool an electrical machine

3. What are the various modes of heat transfer from the inner most part of a machine to the surroundings.

4. Differentiate between natural and artificial convections.

5. Derive an expression for the temperature rise curve for electrical machine.

6. Explain the term heating time constant and show that it has small value for well ventilated machine and large value for poorly ventilated machine.

7. Discuss the advantages of hydrogen as cooling medium as compared to air. What special precaution should be taken for hydrogen cooled alternator

8. Explain various types of cooling system for rotating electrical machines.

9. Distinguish between open circuit and closed circuit ventilation.

10. Distinguish between radial and axial ducts.

11. Describe with relevant diagrams the radial ventilating system, axial ventilating system and combined ventilating system for the cooling of electrical machines.

12. Write short note on hydrogen cooling.

DESIGN OF TRANSFORMER

5.1 INTRODUCTION

A transformer consists of two windings coupled through a magnetic medium.

- The two windings work at different voltage level.
- The two windings of the transformer are called High voltage winding and low voltage winding.
- Both the windings are wound on a common core.
- One of the winding is connected to ac supply and it is called primary.
- The other winding is connected to load and it is called secondary.
- The transformer is used to transfer electrical energy from high voltage winding to low voltage winding or vice-versa through magnetic field.
- The construction of transformers varies greatly, depending on their applications, winding voltage and current ratings and operating frequencies.
- The two major types of construction of transformers (used in transmission and distribution of electrical energy) are core type and shell type.
- Depending on the application, these transformers can be classified as distribution transformers and power transformers.
- The transformer is extremely important as a component in many different types of electric circuits, from small-signal electronic circuits to high voltage power transmission systems.

The most important function performed by transformers are,

- Changing voltage and current level in an electric system.
- Matching source and load impedances for maximum power transfer in electronic and control circuitry.
- Electrical isolation.

5.1.1 Construction of Transformer

In construction of transformer consists following parts.

- Transformer core
- Winding
- Insulation
- Tank

- Conservator and breather
- Tapping and tap changing
- Buchholz Relay
- Explosion vent
- Transformer oil

5.1.2 Specification

- Output-kVA
- Voltage-V1/V2 with or without tap changers and tapings
- Frequency-f Hz
- Number of phases – One or three
- Rating – Continuous or short time
- Cooling – Natural or forced
- Type – Core or shell, power or distribution
- Type of winding connection in case of 3 phase transformers – star-star, star-delta, delta-delta, delta-star with or without grounded neutral
- Efficiency, per unit impedance, location (i.e., indoor, pole or platform mounting etc.),temperature rise etc.,

5.2 CLASSIFICATION OF TRANSFORMER

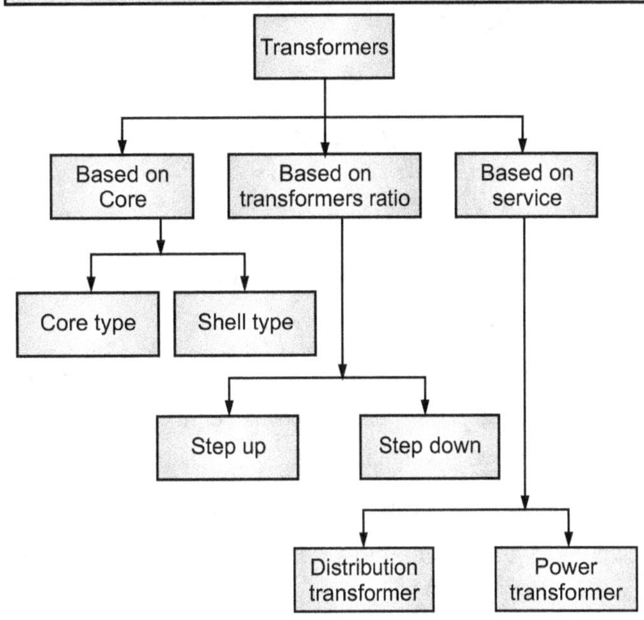

Chart : Classification of transformer

5.2.1 Comparison between Core and Shell Type

1. Core Type Transformer

- In core type transformer, the magnetic core is built of laminations to form a rectangular frame and the windings are arranged concentrically with each other around the legs or limbs.

- The top and bottom horizontal portion of the core are called yoke.

- The yokes connect the two limbs and have a cross sectional area equal to or greater than that of limbs.

- Each limb carries one half of primary and secondary.

- The two windings are closely coupled together to reduce the leakage reactance.

- The low voltage winding is wound near the core and high voltage winding is wound over low voltage winding away from core in order to reduce the amount of insulating materials required.

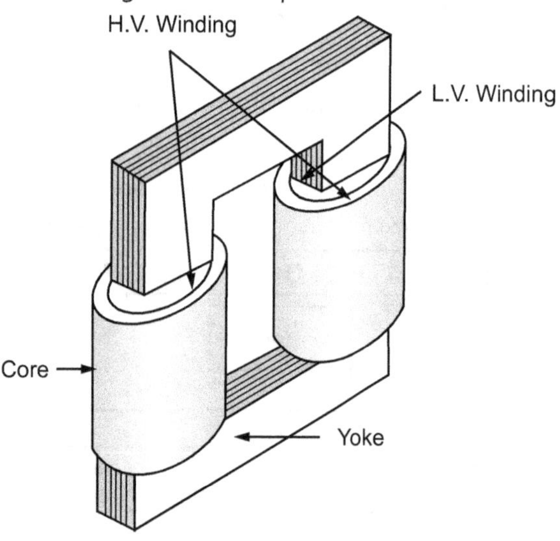

Fig. 5.1 (a) : Core type

2. Shell Type Transformer

- In shell type transformers the windings are put around the central limb and the flux path is completed through two side limbs.

- The central limb carries total mutual flux while the side limbs forming a part of a parallel magnetic circuit carry half the total flux.

- The cross sectional area of the central limb is twice that of each side limbs

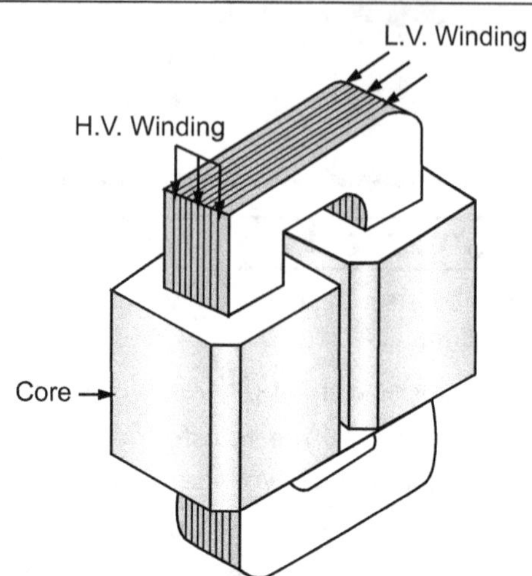

Fig. 5.1 (b) : Shell type

Table 5.1

Sr. No.	Core Type	Shell Type
1.	Easy in design and construction.	Comparatively complex.
2.	Has low mechanical strength due to non-bracing of windings.	Reduction of leakage reactance is highly possible.
3.	Reduction of leakage reactance is not easily possible.	Reduction of leakage reactance is highly possible.
4.	The assembly can be easily dismantled for repair work.	It cannot be easily dismantled for repair work.
5.	Better heat dissipation from windings.	Heat is not easily from windings since it is surrounded by core.
6.	Has longer mean length of core and shorter mean length of coil turn. Hence best suited for EHV (Extra High Voltage) requirements.	It is not suitable for EHV (Extra High Voltage) requirements.

Table 5.2

Description	Core Type	Shell Type
Construction	Easy to assemble and dismantle	Complex
Mechanical strength	Low	High
Leakage reactance	Higher	Smaller
Cooling	Better cooling of winding	Better cooling of core
Repair	Easy	Hard
Application	High voltage and low output	Low voltages and large output

5.2.2 Comparison between Distribution Transformer and Power Transformer

1. Distribution Transformer

- Transformers up to 500kVA are used to step down distribution voltage to a standard service voltage or from transmission voltage to distribution voltage are known as distribution transformers.

- They are kept in operation all the 24 hours a day whether they are carrying any load or not.

- The load on the distribution transformer varies from time to time and the transformer will be on no-load most of the time.

- Hence in distribution transformer the copper loss (which depends on load) will be more when compared to core loss (which occurs as long as transformer is in operation).

- Hence distribution transformers are designed with less iron loss and designed to have the maximum efficiency at a load much lesser than full load.

- Also it should have good regulation to maintain the variation of supply voltage with in limits and so it is designed with small value of leakage reactance.

2. Power Transformer

- The transformers used in sub-stations and generating stations are called power transformers.

- They have ratings above 500kVA. Usually a substation will have number of transformers working in parallel.

- During heavy load periods all the transformers are put in operation and during light load periods some transformers are disconnected.

- Therefore the power transformers should be designed to have maximum efficiency at or near full load.

- Power transformers are designed to have considerably greater leakage reactance that is permissible in distribution transformers in order to limit the fault current.

- In the case of power transformers inherent voltage regulation is less important than the current limiting effect of higher leakage reactance.

Table 5.3

Details	Distribution Transformer	Power Transformer
Capacity	Upto 500 kVA	Above 500 kVA
Voltage rating	11, 22, 33 kV / 440 V	400/33kV; 220/11kV …etc.
Connection	Δ/Y, 3ϕ, 4 wire	Δ/Δ; Δ/Y, 3ϕ, 4 wire
Flux density	Upto 1.5 wb/m^2	Upto 1.7 wb/m^2
Current Density	Upto 2.6 A/mm^2	Upto 3.3 A/mm^2
Load	100% for few Hrs, Part load for some time, No-load for few hrs	Nearly on full load
Ratio of iron loss to Cu loss	1 : 3	1 : 1
Regulation	4 to 9%	6 to 10%
Cooling	Self oil cooled	Forced oil cooled

5.3 DESIGN OF TRANSFORMER

5.3.1 Nomenclature

Let,

ϕ_m = Main flux, Wb

B_m = Maximum flux density, Wb/m^2

δ = Current Density, A/mm^2

A_{gt} = Gross Core area, m^2

A_t = Net Core Area

m^2 = Stacking factor x Gross core area

A_c = Area of Copper in Window, m^2

A_w = Window area, m^2

D = Distance between core cetres, m

d = Diameter of circumscribing circle, m

A_w = Window space factor

f = Frequency, H$_z$

E_t = Emf per turn, V

T_p, T_s = No. of turns in primary and secondary winding respectively

I_p, I_s = Current in primary and secondary winding respectively, A

V_p, V_s = Terminal voltage in primary and secondary winding respectively, V

a_p, a_s = Area of conductor of primary and secondary winding respectively, m^2

I_1 = Mean length of flux path in iron, m

I_m = length of mean turn of transformer winding, m

5.3.2 Size of the Transformer

As the iron area of the leg A_i and the window area A_w = (height of the window H_w × Width of the window W_w) increases the size of the transformer also increases.

The size of the transformer increases as the output of the transformer increases.

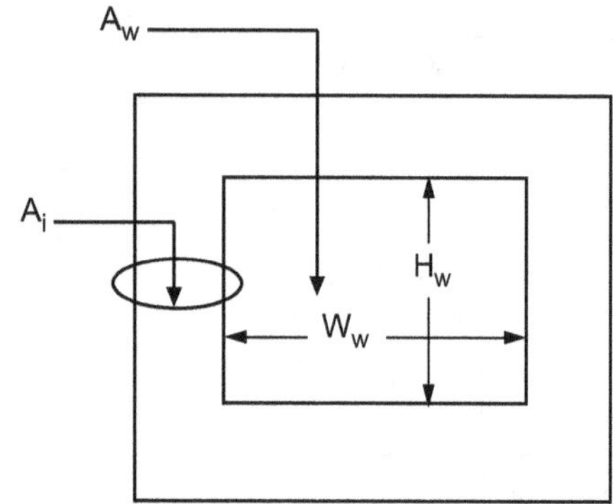

Fig. 5.2 : Size of the Transformer

5.3.3 Output Equation of Transformer

It relates the rated kVA output to the area of core and window.

The output kVA of a transformer depends on,

- Flux Density (B) – related to Core area
- Ampere Turns (AT) – related to Window area
- Window – Space inside the core – to accommodate primary and secondary winding

1. Single Phase Transformer

The Voltage induced in transformer winding is given by

Induced EMF/Turn ,

$$E_t = \frac{E}{T} = 4.44\, f\phi_m \qquad \ldots(5.1)$$

Window in a 1ϕ transformer contains one primary and one secondary winding.

Window space factor,

$$K_w = \frac{\text{Conductor area in window}}{\text{Total area of window}}$$

Window space factor,

$$K_w = \frac{A_c}{A_w}$$

$\therefore$ Conductor area in window

$$A_c = K_w A_w \qquad \ldots(5.2)$$

Current density (δ_e is same in both the windings)

$$\delta = \frac{I_p}{a_p} = \frac{I_s}{a_s} \qquad \ldots(5.3)$$

$\therefore \qquad a_p = \dfrac{I_p}{\delta}\; ;\; a_s = \dfrac{I_s}{\delta}$

If we neglect magnetizing MMF, then

$$(AT)_{primary} = (AT)_{secondary}$$

$$\therefore \qquad AT = I_p T_p = I_s T_s \qquad \ldots(5.4)$$

Total C_u, area in window,

A_c = C_u. area of pry wdg + Cu. area of sec wdg.

$$= \left\{ \begin{array}{c} \text{No. of pry turns X area} \\ \text{of X-section of pry conductor} \end{array} \right\}$$

$$+ \left\{ \begin{array}{c} \text{No. of sec turns X area} \\ \text{of X-section of sec conductor} \end{array} \right\}$$

$$= T_p a_p + T_s a_s$$

$$= T_p \frac{I_p}{\delta} + T_s \frac{I_s}{\delta}$$

$$= \frac{1}{\delta}[T_p I_p + T_s I_s]$$

$$= \frac{1}{\delta}[AT + AT] = \frac{2AT}{\delta} \qquad \ldots(5.5)$$

Therefore, equating (5.2) and (5.5)

$$K_w A_w = \frac{2AT}{\delta}$$

$$AT = \frac{1}{2} K_w A_w \delta \qquad \ldots(5.6)$$

$$Q = V_p I_p \times 10^{-3} \approx E_p I_p \times 10^{-3}$$

$$= \frac{E_p}{T_p} T_p I_p \times 10^{-3} \qquad \left[\text{from (5.1)}, E_t = \frac{E}{T} \right]$$

$$= E_t \cdot AT \times 10^{-3} \qquad \ldots(5.7)$$

$$= 4.44\, f\phi_m \cdot \frac{1}{2} K_w A_w \delta \times 10^{-3}$$

$$= 2.22\, f\phi_m \cdot K_w A_w \delta \times 10^{-3}$$

We know that

$$B_m = \frac{\phi_m}{A_i} \text{ and } \phi_m = B_m A_i$$

$$\therefore \qquad Q = 2.22 \cdot f\, B_m A_i A_w\, K_w\, \delta \times 10^{-3}\ kVA \qquad \ldots(5.8)$$

2. Three Phase Transformer

Total C_u, area in window,

Each window has 2 primary and 2 Secondary windings.

Total Cu. Area in the window is given by,

A_c = C_u. area of pry wdg + Cu. area of sec wdg.

$$= 2 \left\{ \begin{array}{c} \text{No. of pry turns X area} \\ \text{of X-section of pry conductor} \end{array} \right\}$$

$$+ 2 \left\{ \begin{array}{c} \text{No. of sec turns X area} \\ \text{of X-section of sec conductor} \end{array} \right\}$$

$$= 2T_p a_p + 2T_s a_s$$

$$= 2T_p \frac{I_p}{\delta} + 2T_s \frac{I_s}{\delta}$$

$$= \frac{1}{\delta} [2T_p I_p + 2T_s I_s]$$

$$= \frac{1}{\delta} [2AT + 2AT]$$

$$A_c = \frac{4AT}{\delta} \qquad \qquad \ldots(5.9)$$

Compare (5.2) and (5.9)

$$\frac{4AT}{\delta} = K_w A_w$$

$$AT = \frac{K_w A_w \delta}{4}$$

kVA rating of 3ϕ transformer,

$$Q = 3E_p I_p \times 10^{-3}$$

$$= 3 \frac{E_p}{T_p} T_p I_p \times 10^{-3}$$

$$= E_t \times AT \times 10^{-3}$$

$$= 3 \times 4.44 \times f\phi_m \times \frac{1}{4} K_w A_w \delta \times 10^{-3}$$

$$Q = 3.33 \, f B_m A_i A_w K_w \, \delta \times 10^{-3} \text{ kVA} \qquad \ldots(5.10)$$

5.4 EMF PER TURN

Design of transformer starts with the section of EMF/turn.

Let, $\left\{ \begin{array}{c} \text{No. of pry turns X area} \\ \text{of X-section of pry conductor} \end{array} \right\} r = \frac{\phi_m}{AT}$

$$Q = V_p I_p \times 10^{-3}$$

$$= 4.44 \, f\phi_m T_p I_p \times 10^{-3}$$

$$= 4.44 \, f\phi_m (AT) \times 10^{-3}$$

$$= 4.44 \, f\phi_m \frac{\phi_m}{r} \times 10^{-3}$$

$$\phi_m^2 = \frac{Q.r}{4.44 \, f \times 10^{-3}} = \frac{Q \cdot r \times 10^3}{4.44 \, f}$$

$$\phi_m = \sqrt{\frac{Q.r \times 10^3}{4.44 \, f}}$$

w.k.t, $E_t = 4.44 \, f\phi_m$

$$= 4.44 \, f \sqrt{\frac{Q.r \times 10^3}{4.44 \, f}}$$

$$= \sqrt{4.44 \, f} \cdot \sqrt{4.44 \, f} \cdot \sqrt{r \times 10^3} \cdot \frac{\sqrt{Q}}{\sqrt{4.44 \, f}}$$

$$= \sqrt{4.44 \, f \cdot r \times 10^3} \cdot \sqrt{Q}$$

$$E_t = K \cdot \sqrt{Q}$$

where, $K = \sqrt{4.44 f \cdot r \times 10^3} \qquad \ldots(5.11)$

K depends on the type, service condition and method of construction of transformer.

Table 5.4

Transformer Type	Value of K
1ϕ Shell Type	1.0 to 1.2
1ϕ Core Type	0.75 to 0.85
3ϕ Shell Type	1.2 to 1.3
3ϕ Core Type Distribution Xmer	0.45 to 0.5
3ϕ Core Type Power Xmer	0.6 to 0.7

5.5 RATIO OF IRON LOSS TO COPPER LOSS

Ratio of iron loss to copper loss

$$\frac{P_i}{P_c} = \frac{P_i G_i}{P_c G_c}$$

where G_i = weight of active iron,

G_c = weight of copper, kg

P_i = loss in iron per kg, W

P_c = loss in copper per kg, W

The ratio of weight of iron to weight of copper generally lies between 1.5 to 3.0 for distribution transformers.

5.6 OPTIMUM DESIGNS

Transformer may be designed to make one of the following quantitites as minimum.

1. Total Volume
2. Total Weight
3. Total Cost
4. Total Losses

In general, these requirements are contradictory and it is normally possible to satisfy only one of them.

All these quantities vary with

$$r = \frac{\phi_m}{AT}$$

If we take high value of r, flux increases and large are cross-section is needed which increases volume, weight and cost of iron and gives higher loss so value of r is a controlling factor.

5.7 WINDOW SPACE FACTOR

Window space in transformer is fully occupied by conductor material and insulating material. Hence window space factor is defined as the ratio of total copper area in window to total area of window

Window space factor,

$$K_w = \frac{\text{Conductor area in window}}{\text{Total area of window}}$$

Window space factor,

$$K_w = \frac{A_c}{A_w} \angle 1.0$$

Value of Kw, depends on transformer power and voltage rating. Following empirical formula is used for estimating the value of window space factor.

$$K_w = \frac{8}{30 + kV} \quad \text{for transformer rating above 20 kVA}$$

$$K_w = \frac{10}{30 + kV}$$

for transformer rating above 50-200 kVA

$$K_w = \frac{12}{30 + kV}$$

for transformer rating of about 1000 kVA

Where kV is the voltage of h.v. winding in kilo-volt.

5.7.1 Variation in Window Space Factor (Kw) with kVA Rating

Transformers with the same voltage rating but different kVA rating (i) 100 kVA (ii) 1000 kVA, needs same insulating material and large copper material for 1000 kVA compared to 100 kVA. Hence, window space factor value increases as power rating of transformer increases.

5.7.2 Variation in Window Space Factor (Kw) with kV Rating

Transformers with same power rating but different kV rating (i) 11 kV (ii) 110 kV, needs less copper material and more insulating material in 110 kV compared to 11 kV. Hence, window space factor value decreases as voltage level of winding increases.

5.8 SELECTION OF WINDOW DIMENSIONS

- Leakage reactance of transformer depends on distance between adjacent limbs.

- When distance between limbs are small, winding is accommodated by increasing the height i.e. winding is long and thin. This arrangement leads to low value of leakage reactance.

- When distance between limbs are large, winding is accommodated by increasing the width i.e. winding is short and wide. This arrangement leads to high value of leakage reactance.

- The area of window depends upon the total conductor area and window space factor.

Total area of window is defined as

$$A_w = \frac{\text{Total copper area in window}}{\text{Window space factor}}$$

$$= \frac{2a_p T_p}{K_w} \text{ for single phase transformers}$$

$$= \frac{4a_p T_p}{K_w} \text{ for three phase transformers}$$

Area of window A_w = height of window × width of window = $H_w \times W_w$. The ratio of height to width of window, H_w/W_w is between 2 to 4.

Assuming a suitable value for ratio H_w/W_w, the height and width of window can be calculated.

The width of window which gives the maximum output is $W_w = D - d = 0.7\ d$.

5.9 STACKING FACTOR S_F OR K_I

Stacking factor,

$$S_f = \frac{\text{Area of cross-section of iron in the core}}{\text{Area of cross-section of the core}}$$
$$\text{including the insulation area}$$

The usual value of stacking factor is 0.9 & it is denoted by S_f or K_i

5.10 DESIGN OF CORE

- In a transformer, there are low voltage and high voltage windings. The performance of a transformer mainly depends upon the flux linkages between these windings and low reluctance magnetic path is required to link flux between these windings. This low reluctance magnetic path is known as core of transformer.

- Core section of transformer can be square or stepped. These shapes of core can be advantages for circular coils. Distribution and power transformer uses circular coils because mechanical stresses produced at the time of short circuit are radial and hence there is no tendency for the coil to change its shape.

- In small transformer square cores are used. As the size of transformer increase, size of core increase and hence lot of useful space is wasted. Further with the increase in circumscribing circle diameter, length of mean turn of winding increases, which will rise I^2R loss and conductor cost.

- In large transformer, cruciform cores are used. For same core area, space utilization is batter in cruciform cores compared to square core. Further reduction in circumscribing circle diameter, length of mean turn of winding reduces giving low I^2R loss and conductor cost.

5.10.1 Rectangular Core

- It is used for core type distribution transformer and small power transformer for moderate and low voltages and shell type transformers.

- In core type transformer the ratio of depth to width of core varies between 1.4 to 2.

- In shell type transformer width of central limb is 2 to 3 times the depth of core.

$$\frac{\text{Depth}}{\text{Width}} = 1.4 \text{ to } 2$$

Rectangular coils are used.

For shell type,

$$\frac{\text{Width of central limb}}{\text{Depth of Core}} = 2 \text{ to } 3$$

5.10.2 Square and Stepped Core

- Used when circular coils are required for high voltage distribution and power transformer.

- Circular coils are preferred for their better mechanical strength.

- Circle representing the inner surface of the tubular form carrying the windings (Circumscribing Circle)

- Dia of Circumscribing circle is larger in Square core than Stepped core with the same area of cross section.

- Thus the length of mean turn(L_{mt}) is reduced in stepped core and reduces the cost of copper and copper loss.

- However, with large number of steps, a large number of different sizes of laminations are used.

1. Square Core

Let d = diameter of circumscribing circle

Also, d = diagonal of the square core and

           a = side of square

Diameter of circumscribing circle.

$$d = \sqrt{a^2 + a^2} = \sqrt{2a^2} = \sqrt{2}\,a$$

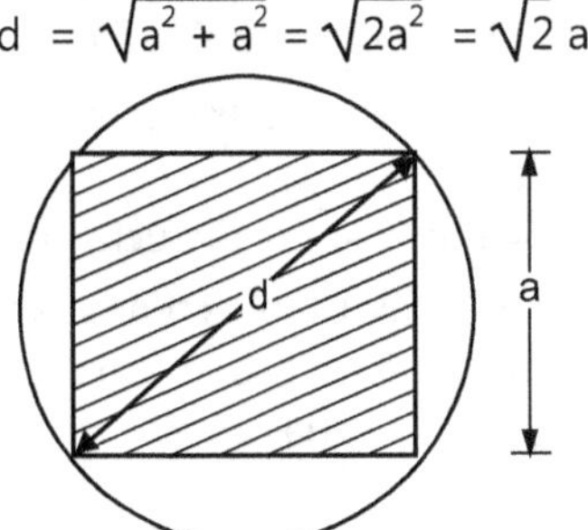

Fig. 5.3 : Square core section

Therefore side of square,

$$a = \frac{d}{\sqrt{2}}$$

Gross core area,

$$A_{gi} = \text{area of square} = a^2$$

$$a^2 = \left(\frac{d}{\sqrt{2}}\right)^2 = 0.5\, d^2$$

Let stacking factor, $S_f = 0.9$

Gross core area includes insulation area

Net core area includes insulation area

$$A_i = 0.9 \times 0.5\, d^2 = 0.45\, d^2$$

Area of circumscribing circle is, $\frac{\pi}{4} d^2$

Ratio of net core area to area of circumscribing circle is

$$\frac{0.45\, d^2}{\pi/r\, d^2} = 0.58$$

Ratio of gross core area to circumscribing circle is

$$\frac{0.5\, d^2}{\pi/4\, d^2} = 0.64$$

Useful ratio in design – Core area factor,

$$K_c = \frac{\text{Net core area}}{\text{Square of circumscribing circle}}$$

$$= \frac{A_i}{d^2} = \frac{0.45\, d^2}{d^2} = 0.45$$

2. Stepped Core or Cruciform Core

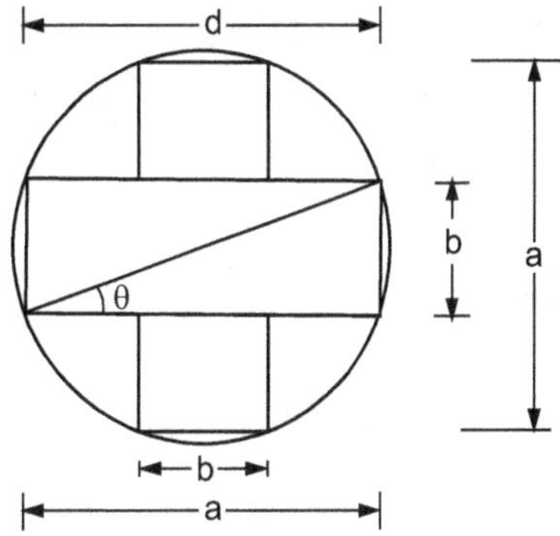

Fig. 5.4 : Cruciform Core

Let, a – Length of the rectangle

            b – breadth of the rectangle

            d – diameter of the circumscribing circle and diagonal of the rectangle.

            θ – Angle b/w the diagonal and length of the rectangle

The max. core area for a given 'd' is obtained by the max value of 'θ'

For max value of 'θ',

$$\frac{dA_{gi}}{d\theta} = 0$$

From the Fig. 5.4

$$\cos \theta = \frac{a}{d} \Rightarrow \therefore a = d \cos \theta$$

$$\sin \theta = \frac{b}{d} \Rightarrow \therefore b = d \cos \theta$$

Two Stepped Core can be Divided in to 3 Rectangles.

Referring to the Fig. 5.5 shown,

Gross core area,

$$A_{gi} = ab + \left(\frac{a-b}{2}\right)b + \left(\frac{a-b}{2}\right)b$$

$$= ab + \frac{2(a-b)}{2}b$$

$$= ab + ab - b^2 = 2ab - b^2$$

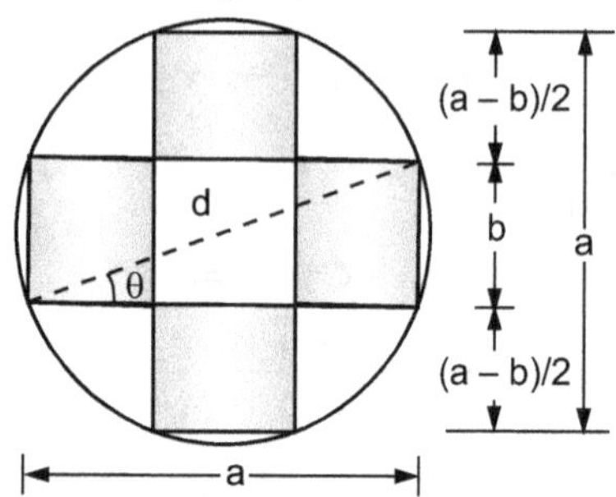

Fig. 5.5: 2-steeped core (cruciform)

On substituting 'a' and 'b' in the above equations.

$$A_{gi} = 2(d \cos \theta)(d \sin \theta) - (d \sin \theta)^2$$

$$A_{gi} = 2d^2 \cos \theta \sin \theta - d^2 \sin^2 \theta$$

$$A_{gi} = d^2 \sin 2\theta - d^2 \sin^2 \theta$$

For max value of 'θ'

$$\frac{dA_{gi}}{d\theta} = 0$$

i.e., $\quad \dfrac{dA_{gi}}{d\theta} = d^2 2 \cos 2\theta - d^2 (2 \sin \theta \cos \theta) = 0$

$$d^2 2 \cos 2\theta = d^2 (2 \sin \theta \cos \theta)$$

$$2 \cos 2\theta = \sin 2\theta$$

$$\frac{\sin 2\theta}{\cos 2\theta} = 2$$

$$\tan 2\theta = 2$$

$$2\theta = \tan^{-1}(2)$$

$$\theta = \frac{1}{2}\tan^{-1}(2) = 31.72°$$

Therefore, if the $\theta = 31.72°$, the dimensions 'a' and 'b' will give maximum area of core for a specified 'd'.

$$\cos \theta = \frac{a}{d} \Rightarrow \therefore a = d \cos \theta$$

$$\Rightarrow \quad a = d \cos(31.72°) = 0.85\ d$$

$$\sin \theta = \frac{b}{d} \Rightarrow \therefore b = d \sin \theta$$

$$\Rightarrow \quad b = d \sin(31.72°) = 0.53\ d$$

As, gross core area,

$$A_{gi} = 2ab - b^2$$

$$A_{gi} = 2(0.85\ d)(0.53\ d) - (0.53\ d)^2$$

$$A_{gi} = 0.618\ d^2$$

Let stacking factor,

$$S_r = 0.9$$

Net core area,

$$A_i = \text{Stacking factor} \times \text{Gross Core area}$$

$$A_i = 0.9 \times 0.618\ d^2 = 0.56\ d^2$$

The ratios

$$\frac{\text{Net core area}}{\text{Area of Circumscribing circle}} = \frac{0.56\ d^2}{\frac{\pi}{4}d^2} = 0.71$$

$$\frac{\text{Gross core area}}{\text{Area of circumscribing circle}} = \frac{0.618\ d^2}{\frac{\pi}{4}d^2} = 0.79$$

Core area factor,

$$K_c = \frac{\text{Net core area}}{\text{Square of circumscribing circle}}$$

$$= \frac{A_i}{d^2} = \frac{0.56\ d^2}{d^2} = 0.56$$

Table 5.5: Ratio of Multi-Stepped Cores

Ratio	Square core	Cruciform core	3-Stepped core	4-Stepped Core
$\dfrac{\text{Net core area}}{\text{Area of circumscribing circle}}$	0.64	0.79	0.84	0.87
$\dfrac{\text{Gross core area}}{\text{Area of circumscribing circle}}$	0.58	0.71	0.75	0.78
Core area factor, K_c	0.45	0.56	0.6	0.62

5.11 CHOICE OF FLUX DENSITY

- Value of flux density in the transformer determines the core area and yoke area, hence computation of flux density is very important and crucial part in design.
- Normally flux density is chosen near knee point of the magnetization curve with some margin to overcome over fluxing, voltage variation and frequency variation.
- Magnetic material used for core and yoke of transformer are hot rolled silicon steel and cold rolled grain oriented silicon steel.

- Choice of flux density may affect the performance parameter such as no load current, behavior under short circuit, iron loss, efficiency and temperature rise.
- Higher value of flux density results in reduced core area and hence there is a saving in iron. With the reduction in core area, the length of mean turn of winding gets reduced which further saves conductor material. Lesser iron and copper material brings down overall cost, weight and size of transformer.
- But higher value of flux density increases the iron loss which results in low efficiency. Increased iron loss causes high temperature rise in core.
- Flux density to be chosen depends on the service condition of the transformer. For distribution transformer high all day efficiency is main design aspect and hence low value of flux density is chosen which keeps down iron loss.

The usual value of maximum flux density for,

Hot rolled silicon steel core material are:

- Power transformer: 1.25 to 1.45 Wb/m^2
- Distribution transformer: 1.10 to 1.35 Wb/m^2

Cold rolled grain oriented silicon steel core material are

- Transformer up to 132 kV: 1.55 Wb/m^2
- Transformer above 275 kV: 1.60 Wb/m^2
- Transformer above 400 kV: 1.70 - 1.75 Wb/m^2

5.12 CHOICE OF CURRENT DENSITY

- The conductor in low voltage and high voltage winding is determined after choosing suitable value of current density.
- Temperature rise may be high if higher value of current density is selected.
- Current density selection is significant for I^2R loss, hence the load at which maximum efficiency occurs depends on it.
- The level of I^2R loss required is different in distribution and power transformer. Thus the value of current density is different for different type of transformer.

  Self-cooled transformer: 1.1 – 2.3 A/mm^2

  Forced air cooled transformer: 2.2 – 3.2 A/mm^2

  Forced oil cooled transformer: 5.4 – 6.2 A/mm^2

5.13 SELECTION OF CORE AREA AND TYPE OF CORE

- Stepped core cross-section is preferred to obtain the optimum core area with the circumscribing circle of the core. The core area is determined by the number of steps, grade of steel insulation or laminations and type of clamping.
- With increases of number of steps core area increases but cost increases. For larger rated transformers high tensile strength clamps are used for clamping core lamination and it provides increased core are for any fixed core diameter. In order to keep the hot spot temperature within specified limits, sufficient number of ducts are to be provided.

5.13.1 Calculation of Core Area

The voltage per turn is,

$$E_t = K\sqrt{Q}$$

where

$$K = \sqrt{4.44\, f \times \frac{\phi_m}{AT} \times 10^3}$$

Now, flux

$$\phi_m = \frac{E_t}{4.44\, f}$$

The value of flux in the core can be calculated. The area of the core is found out by assuming a suitable value of maximum flux density B_m.

Net core area required

$$A_i = \frac{\phi_m}{B_m}$$

and gross core area

$$A_{gi} = \frac{A_i}{k_i}$$

5.14 DESIGN OF WINDING

Transformer windings: HV winding and LV winding

5.14.1 Winding Design Involves

- Determination of no. of turns: based on kVA rating and EMF per turn.
- Area of cross section of conductor used: Based on rated current and Current density.
- No. of turns of LV winding is estimated first using given data.
- Then, no. of turns of HV winding is calculated to the voltage rating.

No. of turns in LV winding,

$$T_{LV} = \frac{V_{LV}}{E_t} \text{ (or) } \frac{AT}{L_{LV}}$$

where,

V_{LV} = Rated voltage of LV winding

I_{LV} = Rated current of LV winding

No. of turns in HV winding,

$$T_{HV} = T_{LV} \times \frac{V_{HV}}{V_{LV}}$$

where, V_{HV} = Rated voltage of HV winding

Area of each primary winding $a_p = \dfrac{I_p}{\delta_p}$

Area of each secondary winding $a_s = \dfrac{I_s}{\delta_s}$

The current densities in the two windings should be taken equal in order to have minimum copper losses i.e. $\delta_f = \sigma_s$. In practice, however, the current density in the relatively better cooled outer winding is made 5 percent greater than the inner winding.

5.14.2 Position of Winding Relative to the Core

LV winding is placed on the inner side and HV winding on the outside. This arrangement is used because the potential difference between LV and core[which is at earth potential] is small, there is less likelihood of a fault occurring between the two. Also, with L.V winding placed nearer to the core, the insulation used between core and winding has a small thickness. However, in case H.V. winding is placed around the core, the insulation between the two has to be thick and this makes the length of mean turn large. This is clear from Fig. 5.6. The cost of insulation is also higher with H.V. in the under side as same amount of insulation has to be used between H.V. and L.V. as is used between H.V. and core while if L.V. is on the inner side, the major insulation is only between L.V. and core.

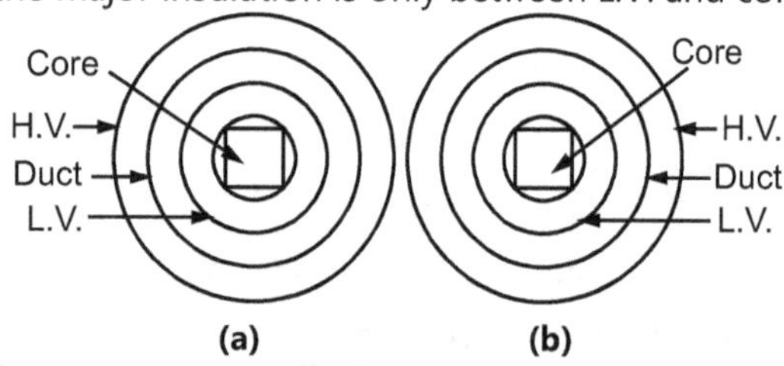

Fig. 5.6: Position of Winding Relative to the Core

SOLVED EXAMPLES

Example 5.1 : *Find that the output of a 3 phase core type transformer is* $Q = 5.23\ fB_m\ Hd^2H_w \times 10^{-3}\ kVA$ *where* f = *frequency, Hz,* B_m = *maximum flux density, Wb/m^2 ;* d = *effective diameter of core, m;* H = *magnetic potential gradient in limb, A/m;* H_w = *height of limb (window), m.*

Solution : kVA output of a three phase transformer

$$Q = 3EI \times 10^3$$
$$= 3 \times 4.44\ f\phi_m\ TI \times 10^{-3}$$
$$= 3 \times 4.44\ f\phi_m\ TI \times 10^{-3}$$

In a three phase core type transformer each limb has one primary and one secondary winding would on it and therefore total mmf over one limb = 2TI.

$\therefore$ Magnetic potential gradient $H = \dfrac{mmf}{\text{height of limb}} = \dfrac{2TI}{H_w}$

or $\qquad TI = \dfrac{HH_w}{2}$ also $A_i = (\pi/4)\ d^2$

Substituting the value of A_i and TI in the expression for Q, we have

$$Q = 3 \times 44\ fB_m \times \frac{\pi}{4}d^2 \times H\frac{H_w}{2} \times 10^{-3}$$

$$Q = 5.23\ fB_m\ Hd^2\ H_w \times 10^{-3}\ kVA$$

Example 5.2 : *Find the core and window areas required for a 1000 kVA, 6600/400 V, 50 Hz, single phase core type transformer. Assume a maximum flux density of 1.25 Wb/m^2 and a current density of 2.5 A/mm^2. Voltage per turn = 30 V. Window space factor = 0.32.*

Given data : kVA = 1000, f = 50 Hz, B_m = 1.25 Wb/m^2

$\qquad V_p$ = 6600 V, V_s = 400 V, δ = 2.5 A/mm^2

$\qquad E_t$ = 30 V, K_w = 0.32, I-phase core type

Solution :

Emf per turn, $E_t = 4.44\ f\phi_m$

$\therefore \qquad \phi_m = \dfrac{E_t}{4.44\ f} = \dfrac{30}{4.44 \times 50} = 0.1351$ Wb

Flux density, $B_m = \dfrac{\phi_m}{A_i}$

$\therefore$ $\left.\begin{array}{l}\text{The net area of} \\ \text{cross-section of core}\end{array}\right\}$ $A_i = \dfrac{\phi_m}{B_m} = \dfrac{0.1351}{1.25}$

$$= 0.108\ m^2 = 0.108 \times 10^6\ mm^2$$
$$A_i = 0.108 \times 10^6\ mm^2$$

The kVA rating or transformer,

$$Q = 2.22\ fB_m A_i K_w\ A_w\ \delta \times 10^{-3}$$

$\therefore$ Window area ,

$$A_w = \frac{Q}{2.22\ fB_m A_i K_w \delta \times 10^{-3}}$$

$$= \frac{1000}{2.22 \times 50 \times 1.25 \times 0.108 \times 0.32 \times 2.5 \times 10^6 \times 10^{-3}}$$

$$= 0.0834\ m^2$$

$$A_w = 0.0834 \times 10^6\ mm^2$$

Example 5.3 : *The ratio of flux to full load mmf in a 400 kVa, 50 Hz, single phase core type power transformer is* 2.4×10^{-6}. *Calculate the net iron area and the window area of the transformer. Maximum flux density in the core is 1.3 Wb/m^2 current density 2.7 A/mm^2 and window space factor 0.26. Also calculate the full load mmf.*

Solution :

$$Q = 400\ kVa$$
$$f = 50\ Hz$$
$$\frac{\phi_m}{AT} = 2.4 \times 10^{-6}$$
$$B_m = 1.3\ wb/m^2$$
$$\delta = 2.7\ A/mm^2$$
$$K_w = 0.26\ \text{Single phase core type}$$

$$K = \sqrt{4.44\, f\, (\phi_m/AT)\, 10^3}$$
$$= \sqrt{4.44 \times 50 \times 24 \times 10^{-6} \times 10^3} = 0.732$$

Voltage per turn

$$E_t = K\sqrt{Q} = 0.732\sqrt{400} = 14.64\ V$$

$$\therefore \quad \text{Flux} \quad \phi_m = \frac{E_t}{4.44\, f} = \frac{14.64}{4.44 \times 50} = 0.066\ Wb$$

Net iron area $A_i = \dfrac{\phi_m}{B_m} = \dfrac{0.066}{1.3} = 0.0507\ m^2$

$$A_i = 0.0507\ m^2$$

Window area of single phase transformer

$$A_w = \frac{Q}{2.22\, fB_m\, K_w\, \delta A_i \times 10^{-2}}$$

$$= \frac{400}{2.22 \times 50 \times 1.3 \times 0.26 \times 2.7 \times 10^6 \times 0.0507 \times 10^{-3}}$$

$$= 0.0777\ m^2$$

$$A_w = 0.0777\ m^2$$

Full load mmf

$$AT = \frac{\phi_m}{2.4 \times 10^{-6}} = \frac{0.066}{2.4 \times 10^{-6}} = 27500\ A$$

$$\mathbf{AT = 27500\ A}$$

5.15 DESIGN OF YOKE

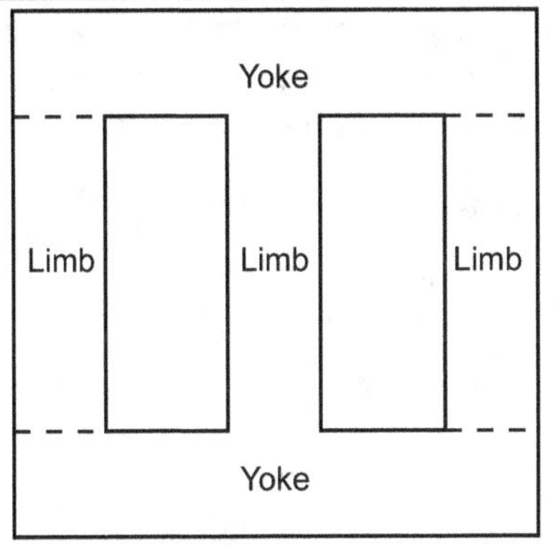

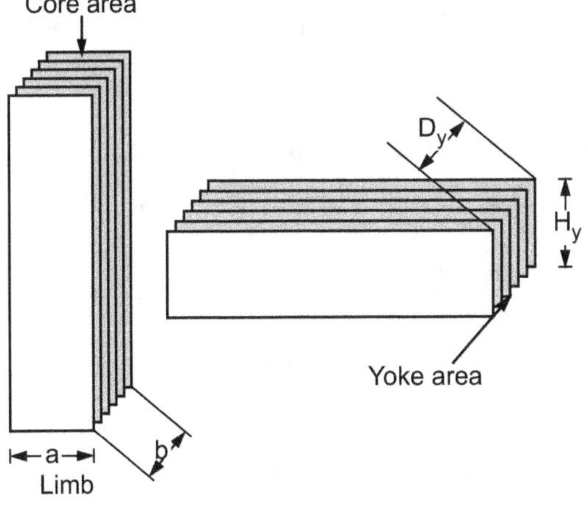

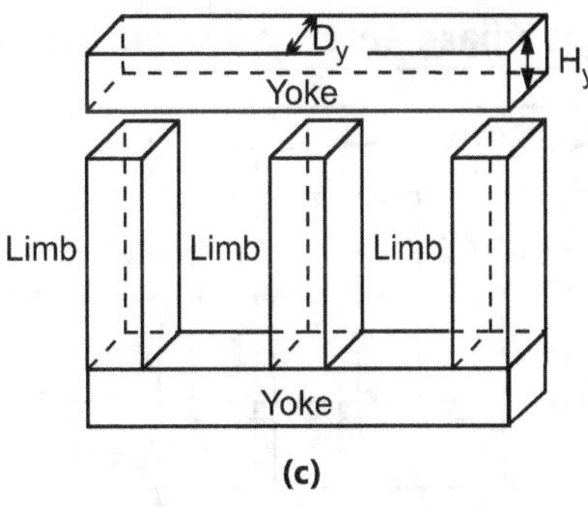

Fig. 5.7 : Yoke Dimensions

- The entire core is divided into two portions. The vertical portion is called core or limb or leg. The horizontal portion is called yoke.

- A yoke is a fixed magnetic part of transformer core which completes the flux path. It is not surrounded by a winding.

- Yoke with rectangular cross section is used for small transformer, however for medium and large transformer two or three stepped yoke is used.

- Number of steps in yoke is much lesser than steps in core hence, unequal distribution of magnetic flux along yoke cross section will give rise to iron loss and no load current in yoke.

- To overcome above said difficulty, cross section area of yoke is taken 10 to 20 % more than cross section area of core.

For rectangular section yokes,

Area of yoke AY = depth of yoke × height of yoke

$$= D_Y \times H_Y$$

where, D_Y = width of largest core stamping

$$= a$$

A_Y = (1.15 to 1.25) A_{gi} for transformers using hot rolled steel.

$= A_{gi}$ for transformers using grain oriented steel.

5.16 OVERALL DIMENSIONS

When dealing with overall dimensions in transformer problems, refer to the following details and diagrams :

a = width of largest stamping

d = diameter of circumscribing circle

D = distance between centres of adjacent limbs

W_w = width of window

H_w = Height of window

   = length of limb

H_y = height of yoke

H = Overall height of transformer over yokes or overall height of frame

5.16.1 Single Phase Core Type Transformer

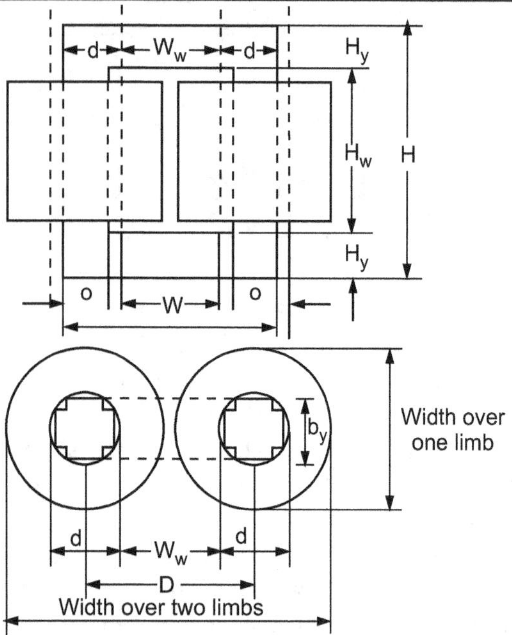

Fig. 5.8: single phase core type transformer

We have the following relations for single phase core type transformers

$$D = d + W_w , \; D_y = a,$$

$$H = H_w + 2H_y, \; W = D + a$$

Width over one limb = outer diameter of h.v. widning

Width over two limbs = D + outer diameter of h.v. winding.

5.16.2 Three Phase Core Type Transformer

We have, for a 3 phase core type transformers

$$D = d + W_w ; \; D_y = a ; \; H = H_w + 2H_y ;$$

$$W = 2D + a$$

Width over one limb = outer diameter of h.v. widning

Width over 3 limbs = 2D + outer diameter of h.v. winding.

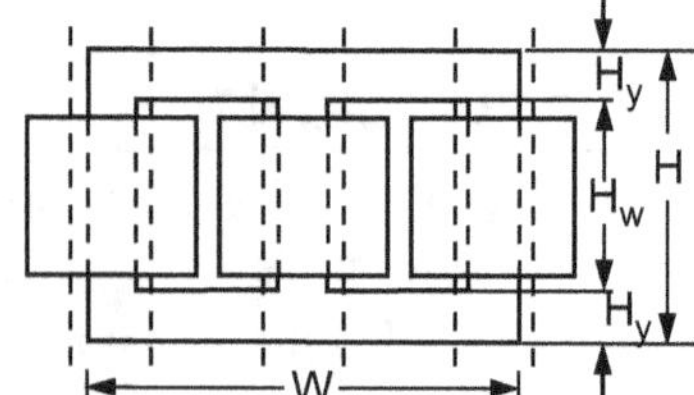

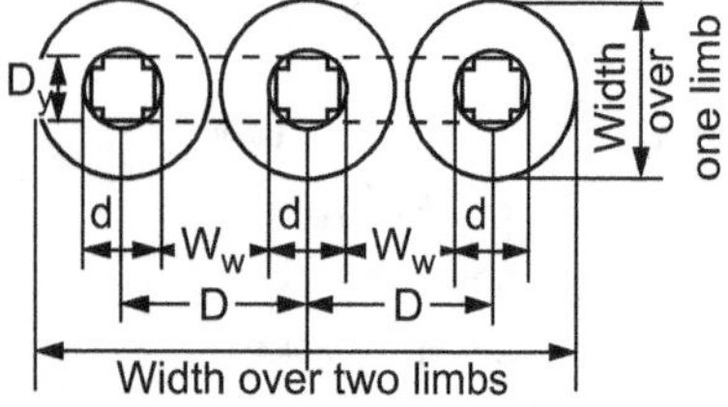

Fig. 5.9: Three phase core type transformer

5.16.3 Single Phase Shell Type Transformer

For single phase shell type referring to Fig. 5.10.

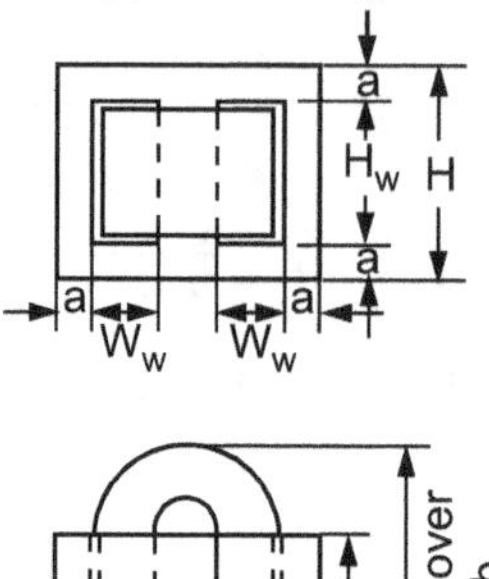

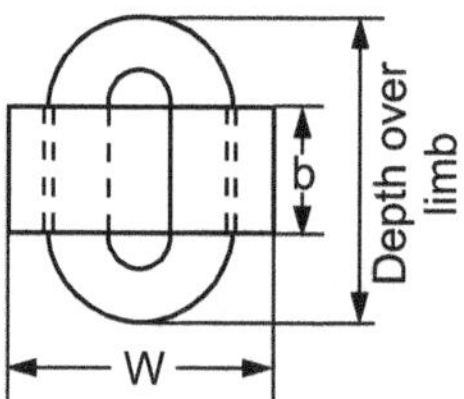

Fig. 5.10: Single Phase shell type transformer

$$D_y = b , \; H_y = a, \; W = 2W_w + 4a , \; H = H_w + 2a$$

Example 5.4 : *Find that the output of a 3 phase core type transformer is : $Q = 5.23 \, f \, B_m \, Hd^2 \, H_w \times 10^{-3}$ kVA where f = frequency, Hz : B_m = maximum flux density, W_b / m^2 ; d = effective diameter of core, m; H = magnetic potential gradient in limb. A/m ; H_w = height of limb (window), m.*

Solution :

kVA output of a three phase transformer $Q = 3EI \times 10^{-3}$

In a three phase core type transformer each limb has one primary and one secondary winding wound on it and therefore total mmf over one limb = 2TI.

$\therefore$　Magnetic potential gradient $H = \dfrac{mmf}{height \; of \; limb}$

$$= \frac{2TI}{H_w} \; \text{or} \; TI = \frac{HH_w}{2} . \; \text{Also} \; A_i = (\pi/4) \, d^2$$

Substituting the value of A_i and TI on the expression for Q, we have

$$Q = 3 \times 4.44 \, f \, B_m \times \frac{\pi}{4} d^2 \times H \frac{H_w}{2} \times 10^{-3}$$

$$\mathbf{Q = 5.23 \, f \, B_m \, Hd^2 H_w \times 10^{-3} \; kVA}$$

Example 5.5 : *Find the kVA output of a single phase transformer from the following data :*

$$\frac{core \; height}{distance \; between \; core \; centres} = 2.8 ,$$

$$\frac{diameter \; of \; circumscribing \; circle}{distance \; between \; core \; centres} = 0.56,$$

$$\frac{net \; iron \; area}{area \; of \; circumscribing \; circle} = 0.7,$$

current density = 2.3 A/mm^2,

window space factor = 0.27,

frequency = 50 Hz

flux density of core = 1.2 Wb/m^2

distance between core centres = 0.4 m

Solution : Given data :

Q = 200 kVA, f = 50 Hz, D = 1.6 a, E_t = 14 V,

B_m = 1.1 wb/m^2, K_w = 0.32, δ = 3 A/mm^2, k_i = 0.9,

A_i = 0.56 d^2, a = 0.85 d

Distance between core centres D = 0.4 m

∴   Core height (window height) H_w = 2.8 × 0.4 = 1.12 m

Diameter of circumscribing circle d = 0.56 × 0.4 = 0.224 m

Width of window W_w = D − d = 0.4 − 0.224 = 0.176 m

∴ Area of window A_w = $H_w \times W_w$ = 1.12 × 0.176 = 0.197 m^2

Area of circumscribing circle = $(\pi/4)$ d^2 = 0.0394 m^2

∴   Net iron area A_i = 0.7 × 0.0394 = 0.0276 m^2

   for a single phase transformer

$$Q = 2.22\ fB_m K_w \delta A_w A_i \times 10^{-3}\ \text{kVA}$$

$$= 2.22 \times 50 \times 1.2 \times 0.27 \times 2.3 \times 10^6$$

$$\times\ 0.197 \times 0.0276 \times 10^{-3}$$

$$= 450\ \text{kVA}$$

$$Q = 450\ \text{kVA}$$

Example 5.6 : *Find the dimensions of core and yoke for a 200 kVA, 50 Hz single phase core type transformer. A cruciform core is used with distance between adjacent limbs equal to 1.1 W_b/m^2, window space factor 0.32, current density 3 A/mm^2, and stacking factor = 0.9. The net iron area is 0.56 d^2 in a cruciform core where d is the diameter circle. Also the width of largest stamping is 0.85 d.*

Solution : Given data :

Q = 200 kVA, f = 50 Hz, D = 1.6 a, E_t = 14V,

B_m = 1.1 wb/m^2, K_w = 0.32, δ = 3 A/mm^2, K_i = 0.9,

A_i = 0.56 d^2, a = 0.85 d

Voltage per turn E_i = 4.44 fϕ_m = 4.44 f$B_m A_i$

∴   Net iron area $A_i = \dfrac{14}{4.44 \times 50 \times 1.1}$ = 0.0573 m^2

∴   Diameter of circumscribing circle d

$$= \sqrt{\dfrac{A_i}{0.56}} = \sqrt{\dfrac{0.0573}{0.56}} = 0.32\ \text{m}$$

Width of largest stamping a = 0.85 d =0.85×0.32=0.272 m.

Distance between core centres D = 1.6 a (given) = 1.6 × 0.272 = 0.435 m.

Width of window W_w = D − d = 0.435 − 0.32 = 0.115 m

for a single phase transformer,

$$Q = 2.22\ fB_m\ K_w\ \delta\ A_w\ A_i \times 10^{-3}$$

$$200 = 2.22 \times 50 \times 1.1 \times 0.32 \times 3 \times 10^6$$

$$\times\ A_w \times 0.0573 \times 10^{-3}$$

∴   Window area A_w = 0.0298 m^2

∴   Height of window $H_w = \dfrac{0.0298}{0.115}$ = 0.26

Using the same stepped section for the yoke as for core

Depth of yoke D_y = a = 0.272 m and height of yoke H_y = 0.272 m.

Using the same stepped section for the yoke as for core

Depth of yoke D_y = a = 0.272 m and height of yoke H_y = 0.272 m

∴   Overall height of frame H = H_w + 2 H_y = 26 + 2 × 0.272 = 0.804 m

∴   Overall length of frame W = D + a = 43.5 + 0.272 = 0.737 m

---

Example 5.7 : *Find overall dimensions for a 200 kVA, 6600/440 V, 50 Hz, 3 phase core type transformer. The following data may be assumed ; emf per turn = 10 V; maximum flux density = 1.3 Wb/m^2; current density = 2.5 A/mm^2; window space factor = 0.3 overall height = overall width ; stacking factor = 0.9. Use a 3 stepped core.*

For a three stepped core :

Width of largest stamping = 0.9 d, and

Net iron area = 0.6 d^2 where d is the diameter of circumscribing circle.

Solution : Given data :

Q = 200 kVA, V_p = 6600 V, V_s = 440 V, f = 50 Hz,

E_t = 10 V, B_m = 1.3 wb/m^2, δ = 2.5 A/mm^2, K_w = 0.3,

H = W, k_i = 0.9, a = 0.9 d, A_i = 0.6 d^2.

Net iron area $A_i = \dfrac{E_t}{4.44\ f\ B_m} = \dfrac{10}{4.44 \times 50 \times 1.3}$ = 0.0347 m^2

Diameter of circumscribing circle d = $\sqrt{\dfrac{0.0347}{0.6}}$ = 0.24 m

and width of largest stamping a = 0.9 × 0.24 = 0.216 m.

Using a 3 stepped section for the yoke.

   Height of yoke H_y = a = 0.216 m

   Depth of yoke D_y = a = 0.216 m

for a 3 phase transformer,

$$Q = 3.33\ fB_m\ K_w\ A_i \times 10^{-3}$$

or$$\quad 200 = 3.33 \times 50 \times 1.3 \times 0.3 \times 2.5 \times 10^6$$

$$\times\ A_w \times 0.0347 \times 10^{-3}$$

∴   Window area A_w = 0.0355 m^2

or $H_w \times W_w$ = 0.0355 m^2

The given condition is, overall height = overall width or H = W.

$$H = H_w + 2\ H_y = H_w + 2 \times 0.216$$

$$= H_w + 0.432$$

$$W = 2D + a = 2(W_w + d) + a$$
$$= 2 W_w + 0.48 + 0.216$$
$$= 2 W_w + 0.696$$

As $H = W$, we have ;
$$H_w + 0.432 = 2 W_w + 0.696$$
or $$H_w = 2 W_w + 0.264$$
$$\therefore (2W_w + 0.264) W_w = 0.0355$$
or $$2 W_w^2 + 0.264 \ W_w - 0.0355 = 0$$

or Width of window $W_w = 0.083$ m

and height of window $H_w = \dfrac{0.0355}{0.083} = 0.428$ m

Hence the dimensions of core are :

Distance between adjacent care centres $D = W_w + d$
$= 0.323$ m

Overall height $H = H_w + 2 H_y = 0.86$ m

Overall width $W = 2D + a = 0.862$ m

Example 5.8 : *The ratio of flux to full load mmf in a 400 kVA, 50 Hz, single phase core type power transformer is 2.4×10^{-6}. Calculate the net iron area and the window area of the transformer. Maximum flux density in the core is 1.3 Wb/m², current density 2.7 A/mm² and window space factor 0.26. Also calculate the full load mmf.*

Solution : Given data :

$Q = 400$ KVA, $f = 50$ Hz, $\phi_m/AT = 2.4 \times 10^{-6}$

$B_m = 1.3$ wb/m² , $\delta = 2.7$ A/mm², $K_w = 0.26$

single phase core type

$$K = \sqrt{4.44 \ f \ (\phi_m/AT) \ 10^3}$$
$$= \sqrt{4.44 \times 50 \times 24 \times 10^{-6}} = 0.732$$

Voltage per turn $E_t = K \sqrt{Q} = 0.732 \sqrt{400} = 14.64$ V

$\therefore$ Flux $\quad \phi_m = \dfrac{E_t}{4.44 \ f} = \dfrac{14.64}{4.44 \times 50} = 0.066$ Wb

Net iron area $A_i = \dfrac{\phi_m}{B_m} = \dfrac{0.066}{1.3} = 0.0507$ m²

Window area of single phase transformer

$$A_w = \dfrac{Q}{2.22 \ f \ B_m \ K_w \ \delta A_i \times 10^{-2}}$$

$$= \dfrac{400}{2.22 \times 50 \times 1.3 \times 0.26 \times 2.7 \times 10^6 \times 0.0507 \times 10^{-3}}$$

$$= 0.0777 \ m^2$$

Full load mmf

$$AT = \dfrac{\phi_m}{2.4 \times 10^{-6}} = \dfrac{0.066}{2.4 \times 10^{-6}}$$

$$= 27500 \ A$$

Example 5.9 : *Find the main dimensions of the core, the number of turns and the cross-section of the conductors for a 5 kVA, $\dfrac{11000}{400}$ V, 50 Hz, single phase core type distribution transformer. The net conductor area in the window is 0.6 times the net cross-section of iron in the core. Assume a square cross-section for the core, a flux density 1 Wb/m², a current density 1.4 A/mm², and a window space factor 0.2. The height of window is 3 times its width.*

Solution : Given :

$Q = 5$ KVA, $V_p = 11000$ V, $V_s = 400$ V, $f = 50$ Hz,

$A_s = K_w A_w = 0.6 \ A_i$, $B_m = 1$ wb/m², $\delta = 1.4$ A/mm²

$H_w = 3 \ W_w$, $k_w = 0.2$

Net conductor area $= 0.6 \times$ net iron area or
$$K_w A_w = 0.6 \ A_i$$

$\therefore$ Window area
$$A_w = \dfrac{0.6}{K_w} A_i = \dfrac{0.6}{0.2} A_i = 3 A_i$$

For a single phase transformer,
$$Q = 2.22 \ f B_m \ K_w \delta \ A_w \ A_i \times 10^{-3}$$
or $$5 = 2.22 \times 50 \times 1.0 \times 0.2 \times 1.4 \times 10^6$$
$$\times 3 A_i \times A_i \times 10^{-3}$$

or net iron area $A_i = 0.00732$ m² .

Gross iron area $A_{gi} = \dfrac{0.00732}{0.9} = 0.00814$ m²

$\therefore$ Width of core $a = \sqrt{0.00814} = 0.09$ m.

Gross iron area provided $= 0.0081$ m²

Net iron area provided $= 0.00729$ m², Window area $A_w = 3 \times 0.00729 = 0.02187$ m².

Height of window $H_w = 3 \ W_w \ \therefore \ 3 W_w^2 = 0.02187$

But $H_w \times W_w = A_w$

or width of window $W_w = 0.085$ m, and height of window $H_w = 3 \times 0.085 = 0.255$ m.

The yoke has the same gross area as the core. Gross area of yoke $A_y = 0.081$ m²

Depth of yoke $D_y = a = 0.09$ m

$\therefore$ Height of yoke $H_y = \dfrac{0.0081}{0.09} = 0.09$ m

Flux $\quad \phi_m = B_m \ A_i = 1.0 \times 72.9 \times 10^{-3}$
$$= 7.29 \times 10^{-3} \ W_b$$

Voltage per turn
$$E_t = 4.44 \ f\phi_m = 4.44 \times 50 \times 7.29 \times 10^{-3}$$
$$= 1.625 \ V$$

Primary turns $= \dfrac{\text{primary voltage}}{\text{secondary voltge}} \times$ secondary turns

$$= \frac{11000}{400} \times 246 = 6765$$

(The turns of the low voltage should be calculated first and that of high voltage afterwards by using the voltage ratio.)

Secondary turns $T_s = \dfrac{\text{secondary voltage}}{\text{voltage per turn}} = \dfrac{400}{1.625} = 246$

Primary winding current $I_p = \dfrac{5000}{11000} = 0.455$ A

Area of primary winding conductor

$$a_p = \frac{I_p}{\delta} = \frac{0.455}{1.4} = 0.384 \text{ mm}^2$$

Using circular conductors, diameter of primary conductor

$$= \sqrt{0.324 \times 4/\pi} = 0.642 \text{ mm}$$

Secondary winding current

$$I_s = \frac{5000}{400} = 12.5 \text{ A}$$

Area of secondary winding conductor

$$a_s = \frac{I_s}{\delta} = \frac{12.5}{1.4} = 8.93 \text{ mm}^2$$

Using a square conductor 3×3 mm^2

Overall dimensions of core

Distance between core centres

$$D = a + W_w = 0.09 + 0.085 = 0.175 \text{ m}$$

Length of frame

$$W = D + a = 0.175 + 0.09 = 0.265 \text{ m}$$

Height of frame

$$H = H_w + 2 H_y = 0.255 + 2 \times 0.09$$
$$= 0.435 \text{ m}$$

Example 5.10 : *Find the main dimensions and winding details of a 100 kV A $\dfrac{2000}{400}$ volt, 50 Hz, single phase shell type, oil immersed, self cooled transformer. Assume voltage per turn, 10 V flux density in core, 1.1 Wb/m^2, current density 2 A/mm^2 window space factor, 0.33.*

The ratio of window height to window width 3 and ratio of core depth to width of central limb = 2.5. The stacking factor is 0.9.

Solution : Given :

Q = 100 KVA, V_p = 2000 V, V_s = 400 V, f = 50 Hz,

E_t = 10 V, k_i = 0.9, B_m = 1.1 wb/m^2, δ = 2 A/mm^2, K_w = 0.33, H_w/W_w = 3, b/2a = 2.5.

Net iron area $A_i = \dfrac{E_t}{4.44 \, fB_m} = \dfrac{10}{4.44 \times 50 \times 1.1} = 0.441 \text{ m}^2$

Gross iron area

$$A_{gi} = \frac{0.041}{0.9} = 0.0555 \text{ m}^2$$

$$\frac{b}{2a} = 2.5 \text{ (given)},$$

We have, gross iron area

$$A_g = 2a \times b$$

The side limbs carry half of the flux in the central limb. Therefore, the width of side limb is half of the width of central limb. Width of side limb a = 0.0675 m

for a single phase transformer,

$$Q = 2.22 \, fB_m \, K_w \, \delta \, A_t \, A_w \times 10^{-3}$$
$$100 = 2.22 \times 50 \times 1.1 \times 0.33 \times 2 \times 10^6$$
$$\times 0.041 \times A_w \times 10^{-3}$$

We have,

$$H_w \times W_w = 0.0303 \text{ and } \frac{H_w}{W_w} = 3$$

$\therefore$ Window area A_w = 0.0303 m^2

or $\qquad 3W_w^2 = 303 \times 10^{-4}$.

Thus width of window W_w = 0.1 m

Height of window H_w = 0.3 m

height of frame $H = H_w + 2H_y$
$$= 0.3 + 2 \times 0.0675$$
$$= 0.435 \text{ m}$$

length of frame

$$W = 2 W_w + 4a$$
$$= 2 \times 0.1 + 4 \times 0.0675 = 0.47 \text{ m}$$

Depth of frame, b = 0.3375 m

Core depth $\quad$ b = 2.5×0.135 = 0.3475 m

The yoke carries half of the flux in the central limb. Assuming the same flux density in the core as the limb, the area of yoke is equal to half the area of central limb.

Gross area of yoke

$$A_y = \frac{0.04555}{2} = 227.75 \times 10^{-3} \text{ m}^2$$

Depth of yoke

$$D_y = b = 0.3375 \text{ m}$$

$\therefore$ Height of yoke $H_y \dfrac{22.275 \times 10^{-3}}{0.3375} = 0.0675$ m

Windings :

H.V winding turns $\quad T_p = \dfrac{2000}{10} = 200$

L.V winding turns $\quad T_s = \dfrac{400}{10} = 400$

H.V winding current $I_p = \dfrac{100 \times 1000}{10} = 50$ A

H.V. winding conductor area

$$a_p = \frac{50}{2} = 25 \text{ mm}^2$$

L.V. winding current

$$I_s = \frac{100 \times 1000}{400} \times 250 \text{ A}$$

L.V. winding area $a_s = \dfrac{250}{2} = 125 \text{ mm}^2$

L.V. winding area $a_s = \dfrac{250}{2} = 125 \text{ mm}^2$

Example 5.11 : *Find out the core window areas required for a 1000 kVA, $\dfrac{6600}{400}$ V, 50 Hz, single phase core type transformer. Assume a maximum flux density of 1.25 Wb/m² and a current density of 2.5 A/mm². Voltage per turn = 30 V. Window space factor = 0.32.*

Solution : Given :

kVA = 1000, f = 50 Hz, B_m = 1.25 Wb/m², 1-phase

V_p = 6600 V, V_s = 400 V, δ = 2.5 A/mm²

E_t = 30 V, K_w = 0.32 core type

Emf per turn, $E_t = 4.44\, f\phi_m$

$\therefore \qquad \phi_m = \dfrac{E_t}{4.44\ f} = \dfrac{30}{4.44 \times 50} = 0.1351$ Wb

Flux density, $B_m = \dfrac{\phi_m}{A_i}$

$\therefore$ The net area of cross-section of core

$$A_i = \frac{\phi_m}{B_m} = \frac{0.1351}{1.25} = 0.108 \text{ m}^2$$

$$= 0.108 \times 10^6 \text{ mm}^2$$

The kVA rating of transformer,

$$Q = 2.22\, f B_m\, A_i\, K_w\, A_w\, \delta \times 10^{-3}$$

$\therefore$ Window area,

$$A_w = \frac{Q}{2.22\, f B_m A_i K_w \delta \times 10^{-3}}$$

$$= \frac{1000}{2.22 \times 50 \times 1.25 \times 0.108 \times 0.32 \times 2.5 \times 10^6 \times 10^{-3}}$$

$$= 0.0834 \text{ m}^2$$

$$= 0.0834 \times 10^6 \text{ mm}^2$$

$\therefore$ Net core area,

$$A_i = 0.108 \text{ m}^2$$

$$= 0.108 \times 10^6 \text{ mm}^2$$

Window area, A_w = 0.0834 m²

$$= 0.0834 \times 10^6 \text{ mm}^2$$

Example 5.12 : *Find the main dimensions including winding conductor area of a 3-phase, Δ-Y core type transformer rated at 300 kVA, $\dfrac{6600}{440}$ V, 50 Hz. A suitable core with 3-steps having a circumscribing circle of 0.25 m diameter and a leg spacing of 0.4 m is available. Emf per turn = 8.5 V, δ = 2.5 A/mm², K_w = 0.28, S_f = 0.9 (stacking factor).*

Solution : Given

3-phase , Δ-Y 50 Hz E_t = 8.5 V, leg spacing = 0.4 m

3-stepped core δ = 2.5 A/mm², Core type S_f = 0.9

300 kVA, d = 0.25, K_w = 0.28 $\dfrac{6600}{440}$ V

Let 440 V side be secondary and 6600 V be primary. Here the secondary is star connected and primary is delta connected.

$\therefore$ Secondary voltage per phase,

$$V_s = \frac{440}{\sqrt{3}} = 254 \text{ V}$$

Also, $E_s \approx V_s$

Emf per turn, $E_t = \dfrac{E_s}{T_s}$

$\therefore$ Number of secondary turns per phase,

$$T_s = \frac{E_s}{E_t} = \frac{259}{8.5} = 29.88 \approx 30 \text{ turns}$$

The phase voltage ratio of transformer,

$$\frac{V_s}{V_p} = \frac{254}{6600}$$

$\therefore$ Number of primary turns per phase

$$T_p = T_s \times \frac{V_p}{V_s} = 30 \times \frac{6600}{254}$$

$$= 779.5 \approx 780 \text{ turns}$$

The kVA rating of transformer,

$$Q = \sqrt{3}\ V_{LP}\, I_{LP} \times 10^{-3}$$

$$= \sqrt{3}\ V_{LS} I_{LS} \times 10^{-3}$$

where V_{LP} = Line voltage on primary side

    I_{LP} = Line current on primary side

    V_{LS} = Line voltage on secondary side

    I_{LS} = Line current on secondary side

Line current on primary side,

$$I_{LP} = \frac{Q}{\sqrt{3}\ V_{LP} \times 10^{-3}}$$

$$= \frac{300}{\sqrt{3} \times 6600 \times 10^{-3}} = 26.24 \text{ A}$$

Since primary is delta connected,

The phase current on primary,

$$I_p = \frac{I_p}{\sqrt{3}} = \frac{26.24}{\sqrt{3}} = 15.15 \text{ A}$$

The area of cross-section of primary conductor

$$a_p = \frac{I_p}{\delta} = \frac{15.15}{2.5} = 6.06 \text{ mm}^2$$

The line current on secondary sides,

$$I_{LS} = \frac{Q}{\sqrt{3} \times V_{LS} \times 10^{-3}}$$

$$= \frac{300}{\sqrt{4} \times 440 \times 10^{-3}} = 393.65 \text{ A}$$

Since secondary is star connected,

The phase current on secondary

$$I_s = I_{LS} = 393.65 \text{ A}$$

The area of cross-section of secondary conductor

$$a_s = \frac{I_s}{\delta} = \frac{393.65}{2.5} = 157.5 \text{ mm}^2$$

The copper area in window

$$A_c = 2\,(a_p T_p + a_s T_s)$$

$$= 2\,(6.06 \times 780 + 157.5 \times 30)$$

$$= 18903.6 \text{ mm}^2$$

Window area, $A_w = \dfrac{A_c}{K_w} = \dfrac{18903.6}{0.28} = 67512.86 \text{ mm}^2$

$$= 67512.86 \times 10^{-6} \text{ m}^2 = 0.0675 \text{ m}^2$$

Area of circumscribing circle $= \dfrac{\pi d^2}{4} = \dfrac{\pi (0.25)^2}{4} = 0.049 \text{ m}^2$

For 3 stepped core, the ratio $\dfrac{\text{Gross core area}}{\text{Area of circumscribing circle}}$

$= 0.84$

Gross core area,

$$A_i = S_f \times A_{gi}$$

$$= 0.9 \times 0.0412 = 0.0369 \text{ m}^2$$

$$= 0.037 \times 10^6 \text{ mm}^2$$

Width of window,

$$W_w = \text{leg spacing} = 0.45 \text{ m}$$

Height of window

$$H_w = \frac{A_w}{W_w} = \frac{0.0675}{0.45} = 0.15 \text{ m}$$

Hence,

Number of primary turns per phase,

$$T_p = 780$$

Number of secondary turns per phase,

$$T_s = 30$$

Area of cross-section of primary conductor,

$$a_p = 6.06 \text{ mm}^2$$

Area of cross-section of secondary conductor,

$$a_p = 157.5 \text{ mm}^2$$

Net core area, $A_i = 0.0369 \text{ m}^2$

Window area, $A_w = 0.0675 \text{ m}^2$

Height of window,

$$H_w = 0.15 \text{ w}$$

Width of window,

$$W_w = 0.45 \text{ m}$$

Example 5.13 : *Find the dimensions of core and window for a 5 kVA, 50 Hz, 1-phase, core type transformer. A rectangular core is used with long side twice as long as short side. The window height is 3 times the width. Voltage per turn = 1.8 V. Space factor 0.2, δ = 1.8 A/mm^2, B_m = 1 Wb/m^2.*

Solution : Given :

$Q = 5$ kVA, Core type $\delta = 1.8$ A/mm^2

$f = 50$ Hz, rectangular core, $B_m = 1$Wb/m^2

I-phase, $E_t = 1.8$ V, long side $= 2 \times$ short side

$H_w = 3\,W_w$, $K_w = 0.2$.

Emf per turn, $E_t = 4.44\, f\phi_m$

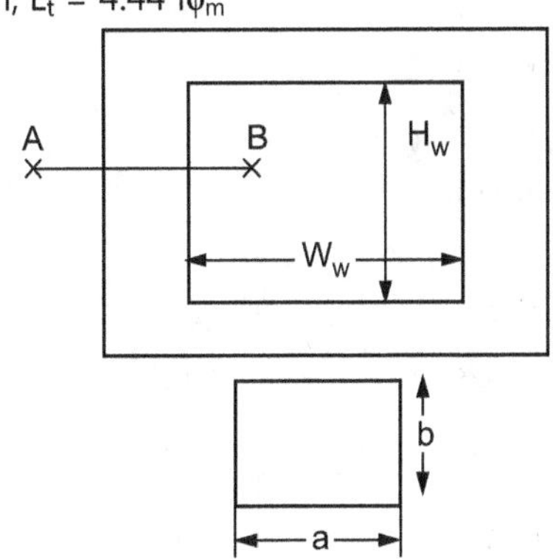

Fig. 5.11

$\therefore \qquad \phi_m = \dfrac{E_t}{4.44\, f} = \dfrac{1.8}{4.44 \times 50} = 0.0081 \text{ Wb}$

Net core area,

$$A_i = \frac{\phi_m}{B_m} = \frac{0.0081}{1} = 0.0081 \text{ m}^2$$

Net core area,

$$A_{gi} = \frac{A_i}{S_f} = \frac{0.0081}{0.9} = 0.009 \text{ m}^2$$

Cross-section of the core is rectangle.

Hence, $\qquad A_{gi} = $ length $\times$ breadth $= a \times b$

Given that $\quad a = 2b$

$\therefore \qquad A_{gi} = 2b \times b = 2b^2$

$$b = \sqrt{\frac{A_{gi}}{2}} = \sqrt{\frac{0.009}{2}} = 0.067 \text{ m}$$

$$a = 2b = 2 \times 0.067 = 0.134 \text{ m}$$

kVA rating of I-phase transformer

$$Q = 2.22 \, fB_m \, A_i \, K_w \, A_w \, \delta \times 10^{-3}$$

Window area,
$$A_w = \frac{Q}{2.22 \, fB_m A_i K_w \delta \times 10^{-3}}$$

$$= \frac{5}{2.22 \times 50 \times 1 \times 0.0081 \times 0.2 \times 1.8 \times 10^6 \times 10^{-3}}$$

$$= 0.0154 \text{ m}^2$$

Also, window area,

$$A_w = H_w W_w$$

Given that $H_w = 3W_w$

Hence, $A_w = H_w W_w = 3W_w \times W_w = 3W_w^2$

$$W_w = \sqrt{\frac{A_w}{3}} = \sqrt{\frac{0.0154}{3}} = 0.0761 \text{ m}$$

$$H_w = 3W_w = 3 \times 0.0716 = 0.2148 \text{ m}$$

Hence,

The Net core area,

$$A_i = 0.0081 \text{ m}^2$$

The dimensions of the core

$$a \times b = 0.134 \times 0.067 \text{ m}$$

The window area,

$$A_w = 0.0154 \text{ m}^2$$

The dimensions of window,

$$H_w \times W_w = 0.2148 \times 0.0716 \text{ m}$$

Example 5.14 : *Find the dimensions of the core, the number of turns, the cross-section area of conductors in primary and secondary windings of a 100 kVA, $\frac{2200}{480}$ V, 1-phase, core type transformer, to operate at a frequency of 50 Hz, by assuming the following data. Approximate volt per turn = 7.5 Volt. Maximum flux density = 1.2 Wb/m². Ratio of effective cross-sectional area of core to square of diameter of circumscribing circle is 0.6. Ratio of height to width is 2. Window space factor = 0.28. Current density = 2.5 A/mm².*

Solution : Given :

100 kVA	50 Hz	$\dfrac{H_w}{W_w} = 2$
$\dfrac{2200}{480}$ V	$E_t = 7.5$ V	I-phase
$K_w = 0.28$	$\dfrac{A_i}{d^2} = 0.6$	$\delta = 2.5$ A/mm²
Core type	$B_m = 1.2$ Wb/m²	

Emf per turn, $E_t = 4.44 \, f\phi_m$

$$\therefore \quad \phi_m = \frac{E_t}{4.44 \, f} = \frac{7.5}{4.44 \times 50} = 0.03378 \text{ W}_b$$

Also, $B_m = \dfrac{\phi_m}{A_i}$

$\therefore$ Net core area

$$A_i = \frac{\phi_m}{B_m} = \frac{0.03378}{1.2} = 0.0282 \text{ m}^2$$

Given that, $\dfrac{A_i}{d^2} = 0.6$. Hence the core is 3-stepped core

$\therefore$ Diameter of circumscribing circle

$$d = \sqrt{\frac{A_i}{0.6}} = \sqrt{\frac{0.0282}{0.6}} = 0.2168 \text{ m}$$

The kVA rating of 1-phase transformer

$$Q = 2.22 \, fB_m A_i K_w \, A_w \delta \times 10^{-3}$$

$\therefore$ Window area,

$$A_w = \frac{Q}{2.22 \, fB_m A_i K_w \delta \times 10^{-3}}$$

$$= \frac{100}{2.22 \times 50 \times 1.2 \times 0.0282 \times 0.28 \times 2.5 \times 10^6 \times 10^{-3}}$$

$$= 0.038 \text{ m}^2$$

$\therefore$ Secondary voltage

$$V_s = 480 \text{ V}$$

Also $E_s \approx V_s$

Emf per turn $= E_t = \dfrac{E_s}{T_s}$

$\therefore$ Number of turns in secondary,

$$T_s = \frac{E_s}{E_t} = \frac{V_s}{E_t} = \frac{480}{7.5} = 64 \text{ turns}$$

The voltage ratio of transformers

$$\frac{V_s}{V_t} = \frac{480}{2200}$$

$\therefore$ Number of turns in primary,

$$T_p = T_s \times \frac{V_p}{V_s} = 64 \times \frac{2200}{480}$$

$$= 293 \text{ turns}$$

The kVA rating of single phase transformer

$$Q = V_p I_p \times 10^{-3} = V_s I_s \times 10^{-3}$$

$\therefore$ Current in primary,

$$I_p = \frac{Q}{V_p \times 10^{-3}} = \frac{100}{2200 \times 10^{-3}}$$

$$= 45.45 \text{ A}$$

Area of cross-section of primary conductor

$$a_p = \frac{I_p}{\delta} = \frac{45.45}{2.5} = 18.1818 \text{ mm}^2$$

Current in secondary,

$$I_s = \frac{Q}{V_s \times 10^{-3}} = \frac{100}{480 \times 10^{-3}} = 208.33 \text{ A}$$

Area of cross section of secondary conductor

$$a_s = \frac{I_s}{\delta} = \frac{208.33}{2.5} = 83.33 \text{ mm}^2$$

Hence,

Net core area, $A_i = 0.0282 \text{ m}^2$

Diameter of circumscribing circle

$$d = 0.2168 \text{ m}$$

Window area, $A_w = 0.038 \text{ m}^2$

Window dimension,

$$H_w \times W_w = 0.2756 \times 0.1378 \text{ m}$$

Number of turns in primary,

$$T_p = 293 \text{ turns}$$

Number of turns in secondary,

$$T_s = 64 \text{ turns}$$

Area of cross-section of primary conductor,

$$a_p = 18.18 \text{ mm}^2$$

Area of cross-section of secondary conductor,

$$a_s = 83.33 \text{ mm}^2$$

Hence,

Net core area,

$$A_i = 0.0282 \text{ m}^2$$

Diameter of circumscribing circle,

$$d = 0.2168 \text{ m}$$

Window area, $A_w = 0.038 \text{ m}^2$

Window dimension

$$H_w \times W_w = 0.2756 \times 0.1378 \text{ m}$$

Number of turns in primary

$$T_p = 293 \text{ turns}$$

Number of turns in secondary

$$T_s = 64 \text{ turns}$$

Area of cross section of primary conductor,

$$a_p = 18.18 \text{ mm}^2$$

Area of cross section of secondary conductor,

$$a_s = 83.33 \text{ mm}^2$$

5.17 RESISTANCE OF WINDING

ρ = resistivity of conductor material (Ω-m)

T_{HV} = Number of high voltage winding turns

T_{LV} = Number of low voltage winding turns

I_{HV} = Phase current in high voltage winding (A)

I_{LV} = Phase current in low voltage winding (A)

a_{HV} = Cross section area of high voltage winding conductor (mm^2)

a_{LV} = Cross section area of low voltage winding conductor (mm^2)

r_{HV} = Resistance of high voltage winding (Ω)

r_{LV} = Resistance of low voltage winding (Ω)

Lmt_{LV} = Length of mean turn of high winding (m)

Lmt_{HV} = $Lmt_{LV} = L_o = Lmt$

Resistance of winding

$$r_{HV} = \rho \frac{(Lmt_{HV})\,(T_{HV})}{a_{HV}}$$

$$r_{LV} = \rho \frac{(Lmt_{LV})\,(T_{LV})}{a_{LV}}$$

Total resistance referred to high voltage winding

$$R_{HV} = r_{HV} + \left(\frac{T_{HV}}{T_{LV}}\right)^2 r_{LV}$$

Per unit resistance

$$\varepsilon_r = \frac{I_{HV} R_{HV}}{V_{HV}}$$

5.18 LEAKAGE REACTANCE OF WINDING

Winding leakage reactance of 3-ϕ core type transformer clearly stating the assumptions used • Assumptions o High voltage and low voltage winding have equal axial length. o Flux path is parallel to the windings along the axial length. o Mmf required to iron path is negligible. o High voltage winding mmf and low voltage winding mmf is equal, hence magnetizing mmf and magnetizing current is zero. o Half of the leakage flux in the duct links with each winding. o Length of mean turn of each winding is equal. o Reluctance of flux path through yoke negligible. o Windings are uniformly distributed, hence winding mmf varies uniformly from zero to AT from one end to another end.

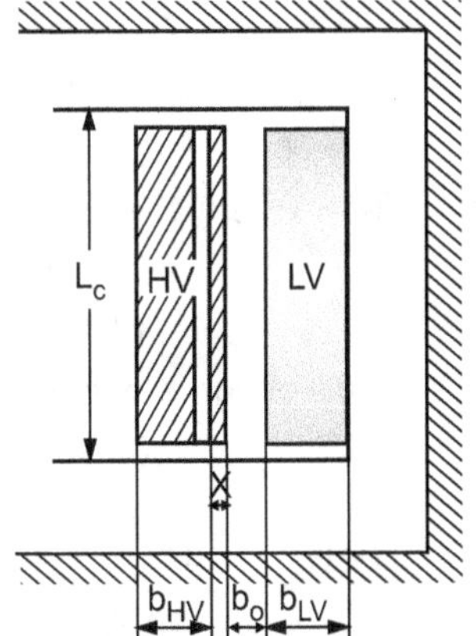

Fig. 5.12: Leakage flux distribution in core type transformer

(a) Conductor Section

Consider a small infinite strip of width dx at a distance x from the edge of high voltage winding along its width.

$$\text{Mmf cross strip} = \left(\frac{x}{b_{HV}}\right)(I_{HV}T_{HV})$$

$$\text{Permeance of strip} = \frac{1}{\text{Reluctance}} = \frac{1}{\dfrac{L_C}{\mu_0 A}} = \frac{\mu_0 (Lmt_{HV}dx)}{L_C}$$

$$= \mu_0 \left(\frac{Lmt_{HV}}{L_C}\right) dx$$

$$\text{Flux in strip} = (\text{Mmf})(\text{permeance})$$

$$= \left[\left(\frac{x}{b_{HV}}\right)(I_{HV}T_{HV})\right]\left[\mu_0 \left(\frac{Lmt_{HV}}{L_C}\right)dx\right]$$

Flux linkages of strip

$$= \left[\left(\frac{x}{b_{HV}}\right)(T_{HV})\right]\left[\mu_0 \left(\frac{Lmt_{HV}}{L_C}\right)(I_{HV}T_{HV})\left(\frac{x}{b_{HV}}\right)\right]dx$$

$$d\psi_1 = \mu_0 \left(\frac{Lmt_{HV}}{L_C}\right)I_{HV}(T_{HV})^2 \left(\frac{x}{b_{HV}}\right)^2 dx$$

Flux linkages of high voltage winding

$$= \int_0^{b_{HV}} d\psi_1$$

$$\psi_1 = \mu_0 \left(\frac{Lmt_{HV}}{L_C}\right)I_{HV}(T_{HV})^2 \int_0^{b_{HV}} \left(\frac{x}{b_{HV}}\right)^2 dx$$

$$\psi_1 = \mu_0 \left(\frac{Lmt_{HV}}{L_C}\right)I_{HV}(T_{HV})^2 \left(\frac{b_{HV}}{3}\right)$$

(b) Duct Section

$$\text{Mmf across duct} = (I_{HV}T_{HV})$$

$$\text{Permeance of duct} = \frac{1}{\text{Reluctance}}$$

$$= \frac{1}{\dfrac{L_C}{\mu_0 A}} = \frac{\mu_0 A}{L_C} = \frac{\mu_0(L_o b_o)}{L_C}$$

$$= \mu_0 \left(\frac{L_o}{L_C}\right) b_o$$

$$\text{Flux in duct} = (\text{Mmf})(\text{Permenace})$$

$$= [(I_{HV}T_{HV})]\left[\mu_0 \left(\frac{L_o}{L_C}\right) b_o\right]$$

Flux linkages of high voltage winding due to duct flux

$$\psi_0 = (\text{Half of flux in duct})(T_{HV})$$

$$= \left[\mu_0 \left(\frac{L_o}{L_C}\right)I_{HV}(T_{HV})^2 \left(\frac{b_o}{2}\right)\right](T_{HV})$$

$$= \mu_0 \left(\frac{L_o}{L_C}\right)I_{HV}(T_{HV})^2 \left(\frac{b_o}{2}\right)$$

Total flux linkages of high voltage winding

$$\psi_{HV} = \psi_1 + \psi_0$$

$$= \mu_0 \left(\frac{Lmt_{HV}}{L_C}\right)I_{HV}(T_{HV})^2 \left(\frac{b_{HV}}{3}\right)$$

$$+ \mu_0 \left(\frac{L_o}{L_C}\right)I_{HV}(T_{HV})^2 \left(\frac{b_o}{2}\right)$$

$$= \mu_0 \left(\frac{I_{HV}(T_{HV})^2}{L_C}\right)\left(Lmt_{HV}\left(\frac{b_{HV}}{3}\right) + L_o\left(\frac{b_o}{2}\right)\right)$$

If, it is assumed that $Lmt_{HV} = Lmt_{LV} = L_o = Lmt$

$$\psi_{HV} = \mu_0 I_{HV}(T_{HV})^2 \frac{Lmt}{L_C}\left(\frac{b_{HV}}{3} + \frac{b_o}{2}\right)$$

Leakage inductance of high voltage winding

$$L_{HV} = \mu_0 (T_{HV})^2 \frac{Lmt}{L_C}\left(\frac{b_{HV}}{3} + \frac{b_o}{2}\right)$$

Leakage reactance of high voltage winding

$$X_{HV} = 2\pi f \mu_0 (T_{HV})^2 \frac{Lmt}{L_C}\left(\frac{b_{HV}}{3} + \frac{b_o}{2}\right)$$

Similarly, leakage reactance of low voltage winding

$$X_{LV} = 2\pi f \mu_0 (T_{LV})^2 \frac{Lmt}{L_C}\left(\frac{b_{LV}}{3} + \frac{b_o}{2}\right)$$

Total leakage reactance referred to high voltage winding

$$X_{HV} = X_{HV} + \left(\frac{T_{HV}}{T_{LV}}\right)^2 X_{LV}$$

5.19 ESTIMATION OF NO-LOAD CURRENT

No-load Current of Transformer:

- Magnetizing Component : Depends on MMF required to establish required flux

- Loss Component : Depends on iron loss

5.19.1 No-Load Current of Single Phase Transformer

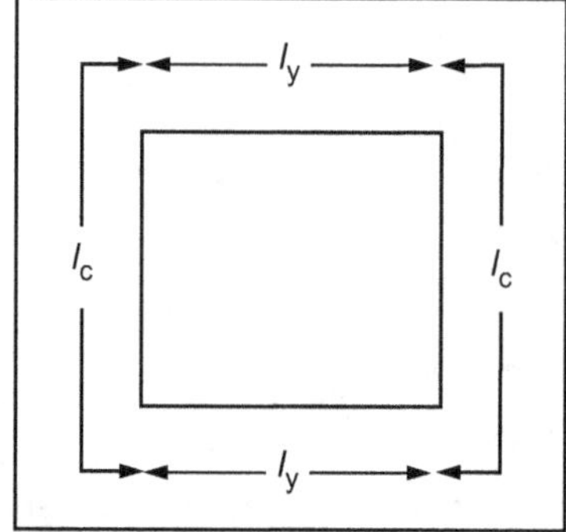

Fig. 5.13

Total Length of the core = $2l_c$

Total Length of the yoke = $2l_y$

Here, $l_c = H_w = $ Height of Window

$$l_y = W_w = \text{Width of Window}$$

MMF for core = MMF per metre for max. flux density in core X Total length of Core

$$= at_c \times 2l_c = 2\, at_c l_c$$

MMF for yoke = MMF per metre for max. flux density in yoke $\times$ Total length of yoke

$$= at_y \times 2l_y = 2\, at_y l_y$$

Total Magnetizing MMF,

$$AT_0 = \text{MMF for Core + MMF for Yoke + MMF for joints}$$

$$= 2\, at_c l_c + 2\, at_y l_y + \text{MMF for joints}$$

The values of at_c and at_y are taken from B-H curve of transformer steel.

Max. value of magnetizing current $= AT_0/T_p$

If the magnetizing current is sinusoidal then,

RMS value of magnetising current, $I_m = AT_0/\sqrt{2}\, T_p$

If the magnetizing current is not sinusoidal,

RMS value of magnetising current, $I_m = AT_0/K_{pk} T_p$

The loss component of no-load current, $I_l = P_i/V_p$

Where, P_i – Iron loss in Watts

V_p – Primary terminal voltage

Iron losses are calculated by finding the weight of cores and yokes. Loss per kg is given by the manufacturer.

No-load current,

$$I_0 = \sqrt{I_m^2 + I_l^2}$$

5.19.2 No-Load Current of Three Phase Transformer

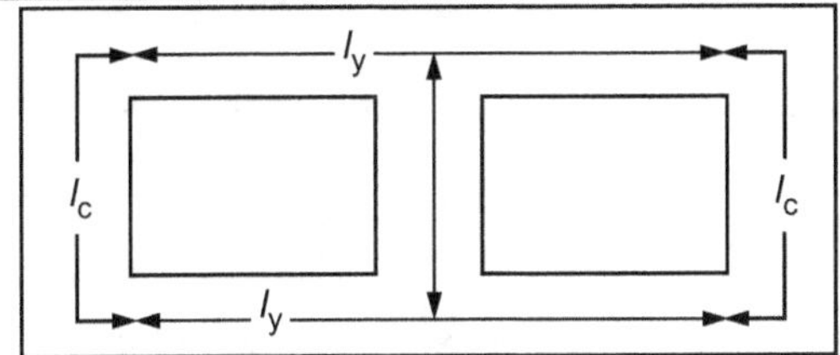

Fig. 5.14

- When transformer operates at no load, current at secondary winding is zero but primary winding draws small current known as no load current.

- No load current of a transformer consists of two components (i) Magnetizing component (I_m) (ii) Loss component (I_1).

Total magnetizing mmf per phase

$$AT_o = \frac{3\,(\text{Mmf of limb}) + 2\,(\text{Mmf of yoke}) + \text{Mmf of joints}}{3}$$

$$AT_o = \frac{3\,at_c l_c + 2\,at_y l_y + \text{Mmf of joints}}{3}$$

Total iron per phase

$$P_i = \frac{\text{Total iron loss}}{3}$$

Magnetizing component of current per phase

$$I_m = \frac{AT_o}{\sqrt{2}\,T_m}$$

Loss components of current per phase

$$I_l = \frac{P_i}{V_{HV}}$$

No load current per phase

$$I_o = \sqrt{(I_m)^2 + (I_l)^2}$$

- Magnetizing component (I_m) produces the magnetic flux in the core, while loss component (I_l) produces real power to feed total iron loss.

- Magnitude of magnetizing component (I_m) depends upon quality of magnetic material used for core and yoke, flux density selected and type of joints.

- Magnitude of loss components (I_l) is very small compared to magnetizing component (I_m). No load current is in order of percentage of rated current of transformer.

Small transformer : 3.0 – 5.0 % of rated current

Medium transformer : 1.0 – 3.0 % of rated current

Large transformer : 0.5 – 2.0 % of rated current

5.20 REGULATION

Fig. 5.15 (a) shows an approximate equivalent circuit of transformer with parameters referred to primary side.

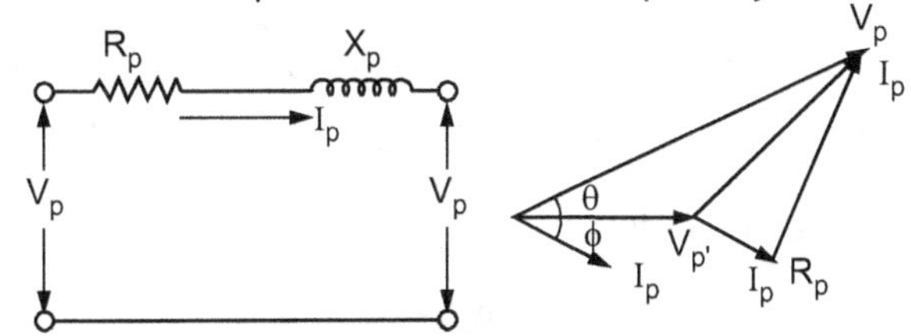

Fig. 5.15 : Simplified equivalent circuit and phase diagram with lagging load

On no load the secondary terminal voltage $V_p' = V_p$. The drop in secondary terminal voltage from no load to full load can be calculated by drawing the phasor diagram. Fig. 5.15 (b) shows the phasor diagram at a lagging power factor $\cos\phi$.

The p.u. regulation, full load rated, for full load rated output Q and full load current I_p is

$$\varepsilon = \frac{V_p - V_p'}{V_p} = \frac{I_p R_p \cos\phi + I_p X_p \sin\phi}{V_p}$$

$$= \varepsilon_r \cos\phi + \varepsilon_p \sin\phi$$

$$\varepsilon = \varepsilon_r \cos\phi + \varepsilon_p \sin\phi$$

Example 5.15 : *Estimate the per unit regulation, at full load and 0.8 power factor aging, for a 300 kVA, 50 Hz, $\dfrac{6600}{400}$ V, 3 phase, delta / star, core type transformer. The data given is :*

H.V. winding : Outside diameter = 0.36 m, insider diameter = 0.29 m, area of conductor = 5.4 mm².

L.V. winding : Outside diameter = 0.26 m, inside diameter = 0.22, area of conductor = 170 mm², Length of coils = 0.5m, voltage per turn = 8v, resistivity = 0.21 Ω/m/mm².

Solution :

L.V. voltage per phase

$$V_s = \frac{400}{\sqrt{3}} = 231 \text{ V}$$

L.V. turns per phase

$$T_s = \frac{V_s}{E_t} = \frac{231}{8} \cong 29.$$

H.V. voltage per phase

$$V_p = 6600 \text{ V}$$

H.V. turns per phase

$$T_p = \frac{6600}{231} \times 29 = 826$$

Mean diameter of l.v. winding

$$= \frac{\text{Outside diameter + inside diameter}}{2}$$

Mean diameter of l.v. winding

$$= \frac{0.26 + 0.22}{2} = 0.24 \text{ m}$$

Length of mean turn of l.v. winding = $\pi \times$ mean diameter

Length of mean turn of l.v. winding = $\pi \times 0.24 = 0.752$ m

Resistance of l.v. winding $= \dfrac{\rho L_{mtp} T_s}{a_p}$

Resistance of l.v. winding $= \dfrac{\rho L_{mts} T_s}{a_s}$

Resistance of l.v. winding

$$r_s = \frac{0.021 \times 29 \times 0.752}{170} = 0.00269 \ \Omega.$$

Mean diameter of h.v. winding

$$= \frac{\text{outside diameter + inside diamter}}{2}$$

Mean diameter of h.v. winding

$$= (0.36 + 0.29)/2 = 0.325 \text{ m}.$$

Length of mean turn of h.v. winding

$$= \pi \times \text{mean diameter}$$

Length of mean turn of h.v. winding

$$= \pi \times 0.325 = 1.02 \text{ m}$$

Resistance of h.v. winding $= \dfrac{\rho \, L_{mtp} T_p}{a_p}$

Resistance of h.v. winding $= \dfrac{0.021 \times 826 \times 1.02}{5.4} = 3.28 \ \Omega$

Resistance of transformer referred to primary = $R_E = r + r\left(\dfrac{T^2}{T}\right)$.

Resistance of transformer referred to primary

$$R_p = 3.28 + 0.00269 \, (826/29)^2 = 5.47 \ \Omega$$

H.V. winding current per phase $I_P = \dfrac{\text{kV.A.} \times 1000}{3V}$

H.V. winding current per phase $I_p = \dfrac{300 \times 1000}{3 \times 6600} = 15.1$ A

P.U. resistance $= \dfrac{IR_{PP}}{V_p}$

$\therefore$ P.U. resistance $\varepsilon_r = \dfrac{15.1 \times 5.47}{6600} = 0.0126$

$$\text{Mean diameter} = \frac{\begin{array}{c}\text{outside diameter} \\ \text{of h.v.}\end{array} + \begin{array}{c}\text{inside diameter} \\ \text{of l.v.}\end{array}}{2}$$

Mean diameter $= \dfrac{(0.36 + 0.22)}{2} = 0.29$ m

Length of mean turn = $\pi \times$ mean diameter

Length of mean turn $L_{mt} = \pi \times 0.29 = 0.91$ m

Width of l.v winding $b_s = \dfrac{(0.26 - 0.22)}{2} = 0.02$ m

Width of h.v. winding $b_p = \dfrac{(0.36 - 0.29)}{2} = 0.035$ m,

Width of duct $a = \dfrac{(0.29 - 0.26)}{2} = 0.015$m

Leakage reactance of transformer referred to primary side

$$X_p = \pi f \mu_0 \, T_p^2 \frac{L_{mt}}{L_c} \left(a + \frac{b_p + b_s}{3}\right)$$

$$X_p = 2\pi \times 50 \times 4\pi \times 10^{-7} \times (826)^2 \times \frac{0.91}{0.5}$$

$$\left(0.015 + \frac{0.035 + 0.02}{3}\right)$$

$$= 17.3 \ \Omega$$

P.U. leakage reactance $= \dfrac{IR_{PP}}{V_p}$

P.U, leakage reactance $\varepsilon_x = \dfrac{15.1 \times 1.73}{6600} = 0.0395$

per unit regulation

$$\varepsilon = \varepsilon_r \cos\phi + \varepsilon_x \sin\phi$$

$$= 0.0126 \times 0.8 + 0.0395 \times 0.6 = 0.0338$$

5.21 MECHANICAL FORCES

- Transformer windings under normal operating conditions are subjected to mechanical forces such as

  - Force of attraction due to current flowing in same direction

  - Force of repulsion due to current flowing in opposite direction

- The magnitude of force on conductor is proportional to product of current in conductor and intensity of magnetic field due to neighboring conductor. Under normal condition current is small, so forces are moderate and not noticeable.

- At the time of short circuit at full voltage current may reach to 10-25 times full load current hence, mechanical forces will reach to 100-625 times normal forces.

- For circular coil, forces are radial hence there will not be any tendency to change the shape of coil. While on rectangular coil, forces are perpendicular to conductor that will tend to deform coil in circular form. Thus circular coils are preferred in transformer winding.

- Mechanical forces are present in transformer due to leakage flux interaction, near the windings carrying current.

- The forces can be classified into axial forces and radial forces, based on resolving the leakage flux component direction.

- The leakage flux is represented in Fig. 5.16 (a). The axial leakage field and radial leakage field are represented in Fig. 5.16 (b) and 5.16 (c), respectively.

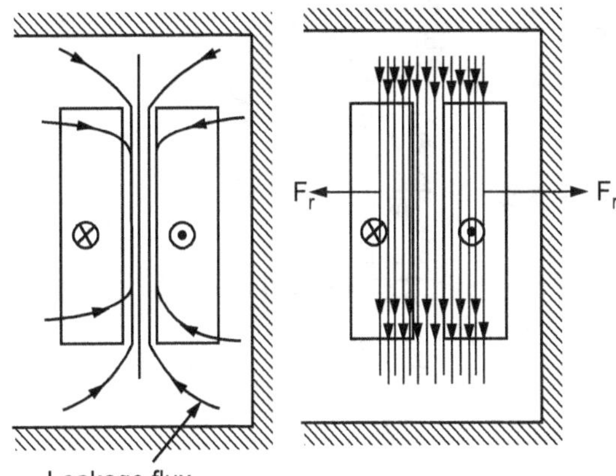

(a) Leakage flux   **(b) Axial leakage field**

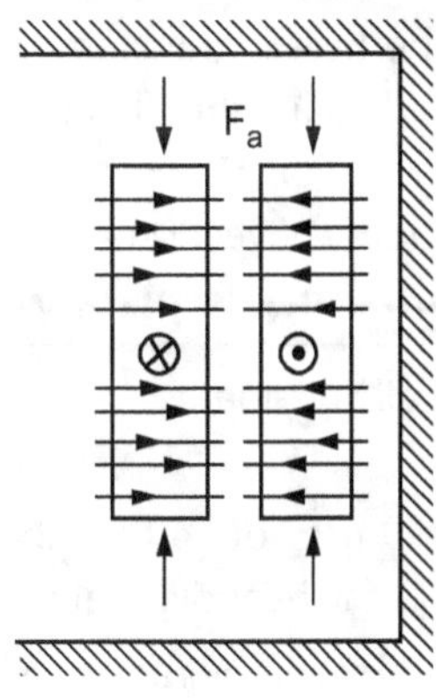

(c) Radial leakage field

Fig. 5.16: Leakage fields and mechanical forces

- Generally, Fleming's left hand rule is used to determine the direction of forces. The cause and effect of the axial and radial forces are represented in Table 5.6

Table 5.6 : Cause and Effect of the Axial and Radial Forces

Feature / Type	Axial Forces	Radial Forces
Nature of occurrence (cause)	• Due to interaction of radial component of leakage flux	• Due to interaction of axial component of leakage flux.
Effect	• compression caused in windings can become tensile if an asymmetry exists in axial magnetic field	• Compression in inner winding. Tensile force in outer winding.

- The magnitude of force acting on a conductor is proportional to the current flow in it and magnetic field intensity due to current in neighbouring conductors. Generally, as the magnetic field is proportional to the current, force is proportional to the square of current.

- The effect of large mechanical forces can be reduced by bracing (reinforcing) of windings.

5.22 TEMPERATURE RISE OF TRANSFORMERS

- Temperature rise in transformer is due to the conversion of loss developed in winding and core into the heat. Heat directed to the cooling medium by the way of conduction, radiation and convection.

- The paths of heat flow are,

  - From internal hot spot to the outer surface(in contact with oil)

  - From outer surface to the oil

  - From the oil to the tank

  - From tank to the cooling medium-Air or water.

- In section (i), the heat is transferred by conduction. In section (ii) and (iii), the heat is transferred by convection of the oil. In section (iv), the heat is transferred by both convection and radiation.

5.22.1 Temperature Rise in Plain Walled Tanks

- Transformer wall dissipates heat in radiation and convection.

For a temperature rise of 40°C above the ambient temperature of 20°C, the heat dissipations are as follows:

- Specific heat dissipation by radiation, $\lambda_{rad} = 6$ W/m$^{2.0}$C

- Specific heat dissipation by convection, $\lambda_{conv} = 6.5$ W/m^2.°C

- Total heat dissipation in plain wall 12.5 W/m^2.°C

The temperature rise,

$$\theta = \frac{\text{Total losses}}{\begin{bmatrix}\text{Specific heat} \\ \text{Dissipation}\end{bmatrix} \times \begin{Bmatrix}\text{heat dissipating} \\ \text{surface of tank}\end{Bmatrix}}$$

$$= \frac{P_i + P_c}{12.5\, S_t}$$

S_t — Heat dissipating surface

- Heat dissipating surface of tank : Total area of vertical sides+ One half area of top cover(Air cooled) (Full area of top cover for oil cooled)

5.23 DESIGN OF TANKS WITH COOLING TUBES

Temperature rise in tank with tubes

- If temperature rise in tank with plain wall exceeds specific limit, additional tubes are attached with the wall to bring down the temperature.

- Additional tube will increase heat dissipation surface area, convection rate but will bring down the radiation rate.

- Hence there is no change by increased surface area so far as dissipation of heat is concern with radiation. But increase in heat dissipation is more effective in convection due to pressure difference created by oil in tubes.

- Let, dissipation surface area increased x times by tank surface area and convection rate by 35%.

  ➢ Length of tube (m)

  ➢ Diameter of tube (m)

  ➢ Area of each tube (m)

  ➢ Number of tube

- Cooling tubes increases the heat dissipation

- Cooling tubes mounted on vertical sides of the transformer would not proportional to increase in area. Because, the tubes prevents the radiation from the tank in screened surfaces.

- But the cooling tubes increase circulation of oil and hence improve the convection

- Circulation is due to effective pressure heads

- Dissipation by convection is equal to that of 35% of tube surface area. i.e., 35% tube area is added to actual tube area.

Let, Dissipating surface of tank – S_t

Dissipating surface of tubes – XS_t

Loss dissipated by surface of the tank by radiation and convection

Loss disspated by tubes by convection } $= 6.5 \times \dfrac{135}{100} \times XS_t$

$$= 8.8\ XS_t$$

Total loss disspated by walls and tubes } $= 12.5\ S_1 + 8.8\ XS_t$

$$= (12.5 + 8.8X)\ S_1 \qquad \qquad ...(5.12)$$

Actual total area of tank walls and tubes

$$= S_t + XS_t = S_t\,(1 + X)$$

Loss dissipated per m^2 of dissipating surface

$$= \frac{\text{Total losses dissipated}}{\text{total area}}$$

Loss dissipated per m^2 of dissipating surface

$$= \frac{S_1\,(12.5 + 8.8\ X)}{S_t\,(1 + X)}$$

$$= \frac{(12.5 + 8.8X)}{(1 + X)} \qquad \qquad ...(5.13)$$

Temperature rise in transformer with cooling tubes

$$\theta = \frac{\text{Total loss}}{\text{Loss dissipated}}$$

Total losses, $P_{loss} = P_i + P_c$

From (5.12) and (5.13), we have

$$\theta = \frac{P_i + P_c}{S_t\,(12.5 + 8.8\ X)}$$

$$(12.5 + 8.8X) = \frac{P_i + P_c}{\theta S_t}$$

$$8.8\ X = \frac{P_i + P_c}{\theta S_t} - 12.5 \Rightarrow$$

$$X = \frac{1}{8.8}\left(\frac{P_i + P_c}{\theta S_t} - 12.5\right)$$

Total area of cooling tubes

$$= \frac{1}{8.8}\left(\frac{P_i + P_c}{\theta S_t} - 12.5\right) S_t$$

$$= \frac{1}{8.8}\left(\frac{P_i + P_c}{\theta} - 12.5\, S_t\right) \qquad \ldots(5.14)$$

Let l_t - length of tubes

d_t - diameter of tubes

$\therefore$ Surface area of tubes $= \pi d_t l_t$

Total number of tubes

$$n_1 = \frac{\text{Total area of tubes}}{\text{Area of each tube}}$$

$$n_1 = \frac{1}{8.8\pi d_1 l_1}\left(\frac{P_i + P_c}{\theta} - 12.5\, S_t\right) \qquad \ldots(5.15)$$

Standard diameter of cooling tube is 50 mm and length depends on the height of the tank.

Centre to centre spacing is 75 mm.

The standard diameter of the cooling tubes is 50 mm and the length of the tube depends on the height of the tank. The tubes are arranged with a centre to centre spacing of 75 mm.

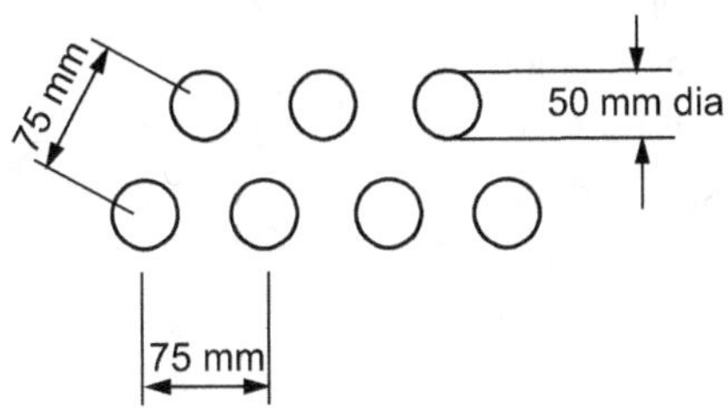

Fig. 5.17 (a)

Dimensions of the Tank

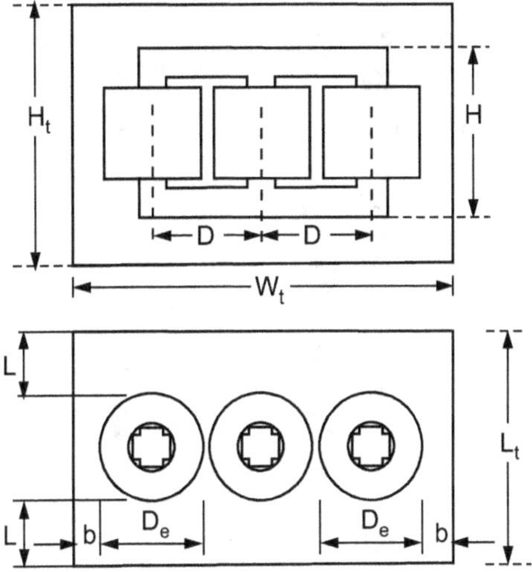

Fig. 5.17 (b) : Tank dimensions

Let,

b – Clearance b/w h.v. winding and tank.

l - Clearance b/w h.v. winding and tank along length

h – Clearance b/w the transformer frame and tank at the bottom and top

D_e – External diameter of the h.v. Winding

Width of the tank,

$$W_T = 2D + D_e + 2b \quad \text{(For 3}\phi\text{ Transformer)}$$
$$= D + D_e + 2b \quad \text{(For 1}\phi\text{ Transformer)}$$

Length of the tank, $L_T = D_e + 2l$

Height of the tank, $H_T = H + h$

- Clearance on the sides depends on the voltage and power ratings.
- Clearance at the top depends on the oil height above the assembled transformer and space for mounting the terminals and tap changer.
- Clearance at the bottom depends on the space required for mounting the frame.
- Typical values of clearance.

Table 5.7

Voltage	kVA Rating	Clearance in mm		
		b	l	h
Up to 11kV	<1000kVA	40	50	450
Upto 11 kV	1000-5000kVA	70	90	420
11kV – 33kV	<1000kVA	75	100	550
11kV – 33kV	1000-5000kVA	85	125	550

Example 5.16 : *A 250 kVA, $\frac{6600}{400}$ V, 3 phase core type transfer has a total of 4800 W at full load. The transformer tank is 1.25 m in height and 1m × 0.5 m in plan. Design a suitable scheme for tubes of the average temperature rise is to be limited to 35 °C. The diameter of tubes is 50 mm and are spaced 75 mm from each other. The average height of tubes is 1.05 m.*

Specific heat dissipation due to radiation and convection is respectively 6 and 6.5 W/m²–°C. Assume that convection is improved by 35 per cent due to provision of tubes.

Solution : Area of plane tank

$$S_t = 2(1 + 0.7) \times 1.25 = 3.75\,\text{m}^2$$

Let the tube area be xS_t.

$\therefore$ Total dissipating surface $= (1 + x) S_i = 3.75 (1 + x)$

Specific loss dissipation $= \dfrac{4800}{3.75(1 + x) \times 35} = \dfrac{36.5}{1 + x}\,\text{W/m}^2\text{-°C}$

Loss dissipated $= \dfrac{12.5 + 8.8x}{1 + x}\,\text{W/m}^2\text{-°C}$

$$\frac{12.5 + 8.8x}{1 + x} = \frac{36.5}{1 \times x} \quad \text{or } x = 2.73$$

$\therefore$ Area of tubes $= 2.73 \times 3.75 = 10.23\,\text{m}^2$

Wall area of each tube $= \pi d_1 t_1 = \pi \times 0.05 \times 1.05 = 0.165\,\text{m}^2$

$\therefore$ Total number of tubes to be provided $= \dfrac{10.23}{0.165} = 0.62$

The tubes are spaced 75 mm apart. Hence, in 1m along the width of tank, we can accommodate 12 tubes leaving 90 mm on each side. In 0.5 m along the depth of tank we can accommodate 5 tubes with 100 mm space on each side. The total tubes provided in the first row along width and depth are $2 \times 12 + 2 \times 5 = 34$. The balance $62 - 34 = 28$ tubes can be provided in second row at the back.

It tubes can be provided in a staggered fashion in each of the two long sides an 4 in each of the two short sides.

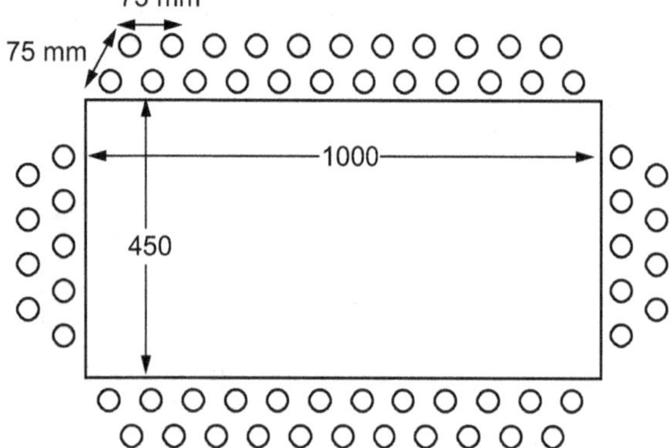

Fig. 5.18 : Arrangement of cooling tubes

$\therefore$ The tubes provident $= 2 \times 12 + 2 \times 11 + 2 \times 5 + 2 \times 4$

$$= 64.$$

Example 5.17 : *The tank of 1250 kVA, natural oil cooled transformer has the dimensions length, width and height as $0.65 \times 1.55 \times 1.85$ m respectively. The full load loss = 13.1 kW, loss dissipation due to radiations = 6 W/m²- ℃, loss dissipation due to convection = 6.5 W/m²- ℃, improvement in convection due to provision of tubes = 40%, temperature rise = 40℃, length of each tube = 1 m, diameter of tube = 50 mm. Find the number of tubes for this transformer. Neglect the top and bottom surface of the tank as regards the cooling.*

Solution : Given :

kVA = 1250, Tank dimension = $0.65 \times 1.55 \times 1.85$ m,

l_t = 1 m, λ_{conv} = 6.5 W/m²-°C, d_t = 50 mm,

λ_{rad} = 6 W/m²-°C, θ = 40°C, Improvement in cooling = 40%, Full load loss = 13.1 kW.

$$L_T = \text{Length} = 0.65 \text{ m}$$

$$W_T = \text{Width} = 1.55 \text{ m}$$

$$H_T = \text{Height} = 1.85 \text{ m}$$

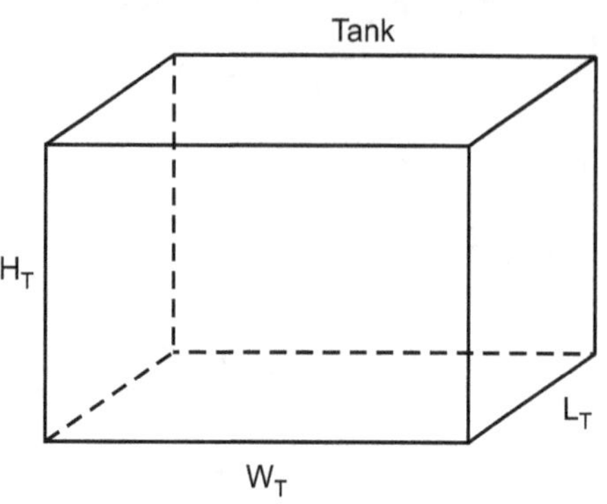

Fig. 5.19

Heat dissipating surface of tank $\Big\}$ S_t = Total Area of Vertical sides

$$= 2\ (L_T H_T + W_T H_T) = 2H_T(L_T + W_T)$$

$$= 2 \times 1.85 \times (0.65 + 1.55) = 8.14 \text{ m}^2$$

Loss dissipating by tank walls by radiation and convection $\Big\}$ $= (6 + 6.5)S_t = 12.5\ S_t$

Let, heat dissipating area of tubes $\Big\}$ $= X\ S_t$

Loss dissipated by cooling tubes due to convection $\Big\}$ $= 6.5 \times \dfrac{140}{100} \times X\ S_t$

$$= 9.1\ X\ S_t$$

Total loss dissipated by tank and tubes $\Big\}$ $= 12.5\ S_t + 9.1\ X\ S_t$

$$= S_t (12.5 + 9.1\ X)$$

Temperature rise in transformer with cooling tubes $\Big\}$ $\theta = \dfrac{\text{Total loss}}{\text{Total loss dissipated}}$

Given that total loss, P_{loss} = 13.1 kW = 13.1×10^3 W

Hence, $\theta = \dfrac{13.1 \times 10^3}{S_t\ (12.5 + 9.1X)}$

$$12.5 + 9.1\ X = \dfrac{13.1 \times 10^3}{\theta\ S_t}$$

$$X = \dfrac{1}{9.1}\left(\dfrac{13.1 \times 10^3}{\theta\ S_t} - 12.5\right)$$

$$= \dfrac{1}{9.1}\left(\dfrac{13.1 \times 10^3}{40 \times 8.14} - 12.5\right) = 3.0476$$

Total area of tubes = $X\ S_t = 3.0476 \times 8.14 = 24.8075 \text{ m}^2$

Total number of cooling tubes $= \dfrac{\text{Total Area of tubes}}{\text{Area each tube}}$

Area if each tube = $\pi\ d_t\ l_t = \pi \times 50 \times 10^{-3} \times 1 = 0.157 \text{ m}^2$

Total number of cooling tubes = $\dfrac{24.8075}{0.157} = 158$ tubes

The diameter of the tube is 50 mm and the standard distance between the tubes is half of the diameter and so, let distance between tubes = 25 mm.

The width of the tank is 1550 mm. If we leave an edge spacing of 62.5 mm on either dies then we can arrange 20 tubes widthwise with a spacing of 75 mm between centres of tubes. On lengthwise we can arrange 8 tubes with same spacing as that of widthwise tubes. But one row is not sufficient to accommodate the required 158 cooling tubes. Hence three rows of cooling tubes are provided on both lengthwise and widthwise. The plan of the cooling tubes is shown in Fig. 5.20.

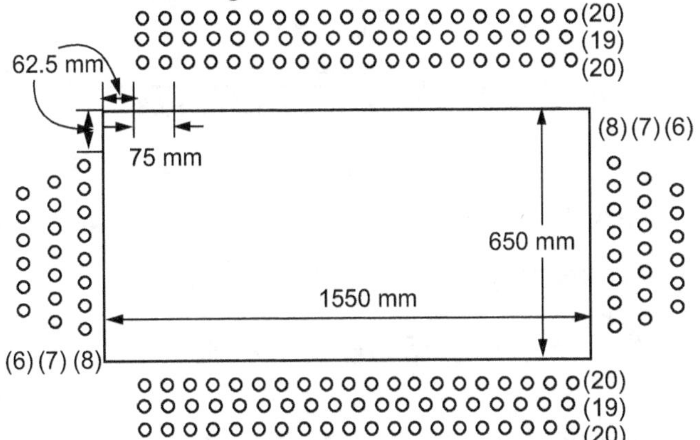

Fig. 5.20 : Plan showing the arrangement of cooling tubes

Result : The total number of tubes provided = 160

They are arranged as 3 row on widthwise with each row consisting of 20, 19, 20 tubes and 3 rows on lengthwise each row consisting of 8, 7 and 6 tubes.

***Example 5.18 :** A 250 kVA, 6600/400 V, 3-phase core type transformer has total loss of 4800 watts on full load. The transformer tank is 1.25 m in height and 1m $\times$ 0.5 m in plan. Design a suitable scheme for cooling tubes if the average temperature rise is to be limited to 35 °C. The diameter of the tube is 50 mm and are spaced 75 mm from each other. The average height of the tube is 1.05 m.*

Specific heat dissipation due to radiation and convection is respectively 6 and 6.5 W/m²- °C. Assume that convection is improved by 35 percent due to provision of tubes.

Solution : Given :

    kVA = 250, Tank dimension = 0.5 × 1 × 1.25 m, θ = 35°C, Total power loss = 4800 W, d_t = 50 mm, Distance between tube centres = 75 mm, l_t = 1.05m, 6600/400V; 3-phase; core type.

$$L_T = \text{Length} = 0.5\text{m};$$
$$W_T = \text{Width} = 1\text{m}$$
$$H_T = \text{Height} = 1.25 \text{ m}$$

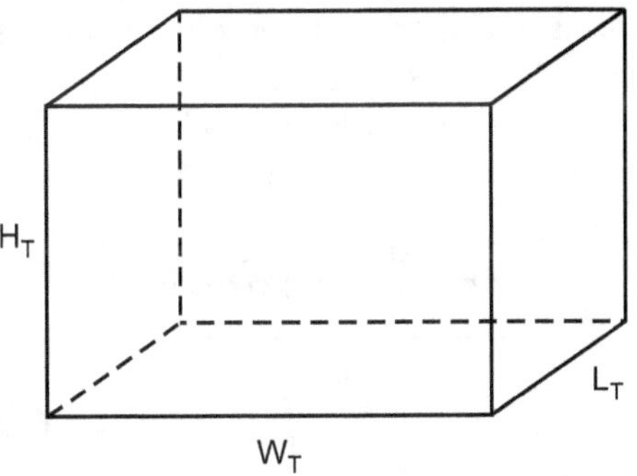

Fig. 5.21

$$\left.\begin{array}{l}\text{Heat dissipating} \\ \text{surface of tank}\end{array}\right\} S_t = \text{Total Area of Vertical sides}$$

$$= 2 (L_T H_T + W_T H_T) = 2H_T(L_T + W_T)$$
$$= 2 \times 1.25 \times (0.5 + 1) = 3.75 \text{ m}^2$$

$$\left.\begin{array}{l}\text{Loss dissipating by tank walls} \\ \text{by radiation and convection}\end{array}\right\} = (6 + 6.5)S_t = 12.5 \; S_t$$

$$\left.\begin{array}{l}\text{Let, heat dissipating} \\ \text{area of tubes}\end{array}\right\} = X \; S_t$$

$$\left.\begin{array}{l}\text{Loss dissipated by cooling} \\ \text{tubes due to convection}\end{array}\right\} = 6.5 \times \frac{135}{100} \times X \; S_t$$

$$= 8.8 \; X \; S_t$$

$$\left.\begin{array}{l}\text{Total loss dissipated} \\ \text{by tank and tubes}\end{array}\right\} = 12.5 \; S_t + 8.8 \; X \; S_t$$

$$= S_t (12.5 + 8.8 \; X)$$

$$\left.\begin{array}{l}\text{Temperature rise in} \\ \text{transformer with cooling tubes}\end{array}\right\} \theta = \frac{\text{Total loss}}{\text{Total loss dissipated}}$$

Given that total loss, P_{loss} = 4800 kW

$$\therefore \quad \theta = \frac{4800}{S_t (12.5 + 8.8X)}$$

Or

$$X = \frac{1}{8.8}\left[\frac{4800}{\theta \; S_t} - 12.5\right]$$
$$= \frac{1}{8.8}\left[\frac{4800}{35 \times 3.75} - 12.5\right] = 2.7354$$

Total area of cooling tubes

$$= X \; S_t = 2.7354 \times 3.75 = 10.2578 \text{ m}^2$$

Area of each cooling tube $= \pi \, d_t \, l_t = \pi \times 50 \times 10^{-3} \times 1.05$
$$= 0.1649 \text{ m}^2$$

Total number of cooling tubes, $n_t = \dfrac{\text{Total Area of tubes}}{\text{Area each tube}}$

$$= \frac{10.2578}{0.1649} = 62.206 \approx 62 \text{ tubes}$$

The width of the tank is 1000 mm. If we leave an edge spacing of 87.5 mm on either sides, then we can arrange 12 tubes widthwise with a spacing of 75 mm between the centres of tubes.

The length of the tank is 500 mm. If we leave an edge spacing of 100 mm on either sides, then we can arrange 5 tubes lengthwise with a spacing of 75 mm between the centres of tubes.

But one row is not sufficient to accommodate the required 62 cooling tubes, Hence 2 rows of cooling tubes are provided on both lengthwise and widthwise. The plan of the cooling tubes is shown in Fig. 5.22.

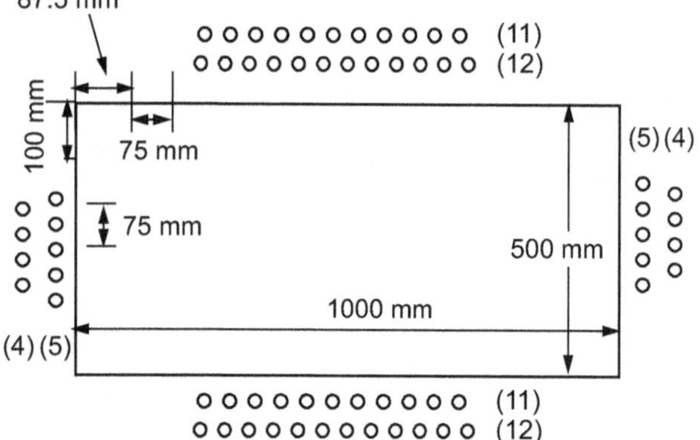

Fig. 5.22 : Plan showing the arrangement of cooling tubes

Result :

Total number of cooling tubes provided = 64

They are arranged as 2 rows on widthwise with each roe consisting of 12 and 11 tubes and 2 rows on lengthwise with each row consisting of 5 and 4 tubes.

Example 5.19 : *A 1000 kV A, 6600/440 V, 50 Hz, 3 phase, delta/star, core type, oil immersed natural cooled (ON) transformer. The design data of the transformer is :*

Distance between centres of adjacent limbs = 0.47 m, outer diameter of high voltage winding = 0.44 m, height of frame = 1.24 m.

Core loss = 3.7 kW and I^2R loss = 10.5 kW.

Solution : Given:

kVA = 1000, V_p = 6600 V, V_s = 440 V, f = 50 Hz, 3 phase, λ_{con} = 6.5 W/m²-°C, D_e = 0.44 m, H = 1.24 m, P_i = 3.7 kW, P_c = 10.5 kW, θ = 35°C, λ_{rad} = 6 W/m² -°C, D = 0.47 m, Improvement in cooling = 35%

Width of tank w_t = 2D + D_e + 2b

Assume   b = 70 mm = 0.07 m

W_t = 2 × 0.47 + 0.44 + 2 × 0.07 = 1.52 m

W_t = 1.52 m

L_t = D_e + 2l

Assume   l = 90 mm = 0.09 m

L_t = 0.44 + 2 × 0.09 = 0.62

L_t = 0.62 m

Height of tank

H_t = H + h

= height of frame + clearance between the assembled transformer and the tank

= 1.24 + (0.05 + 0.3 + 0.3)

H_t = 1.89 m

Dissipating surface of the tank

S_t = 2H_t (W_t + L_t) = 2 × 1.89 (1.52 + 0.62)

S_t = 8.0892 m²

Loss dissipation by transformer tank = (6 + 6.5) S_t

= 12.5 S_t

Loss dissipation by tube = $6.5 \times \dfrac{135}{100} \times X S_t$ = 8.8 X S_t

Total loss dissipation by tank and tube

= 12.5 S_t + 8.8 X S_t

Total loss = P_i + P_c = $3.7 \times 10^3 + 10.5 \times 10^3$

= 14.2×10^3 watts

Temperature rise $\theta = \dfrac{\text{Total loss}}{\text{Loss dissipated}}$

$= \dfrac{P_i + P_c}{12.5\, S_t + 8.8\, X\, S_t}$

$35 = \dfrac{14.2 \times 10^3}{12.5 \times 8.0892 + 8.8 \times x \times 8.0892}$

x = 4.2789

Area of tube = X S_t = 4.2789 × 8.0892 = 34.61 m²

Area of each tube = $\pi\, d_t\, l_t = \pi \times 0.05 \times 1.4$ = 0.2199 m²

Number of tubes = $n_t = \dfrac{\text{Total area of tube}}{\text{Area if each tube}}$

$= \dfrac{34.61}{0.2199}$ = 157.38 ≈ 158

n_t = 158

Total number of cooling tubes provided = 160

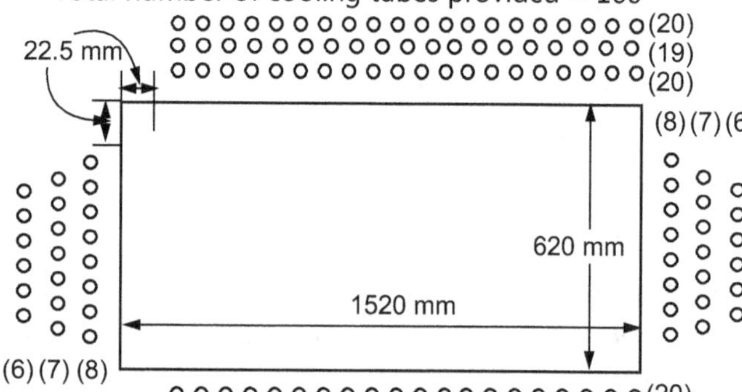

Fig. 5.23 : Plan showing the arrangement of cooling tubes

They are arranged as 3 rows on widthwise with each row consisting of 20, 19 and 20 tubes and 3 rows on lengthwise with each row consisting of 8, 7 and 6 tubes.

5.24 COOLING OF METHODS OF TRANSFORMER

The transformer is static device which converts electrical energy at one voltage level to another voltage level. During this process of energy transfer losses occur in the windings and core of the transformer. These losses appear as heat.

- To dissipate the heat to surroundings air and oil is used as coolant. Transformer using air as coolant is called dry type and oil as coolant is called oil immersed.

- If this heat is not dissipated properly, the excess temperature in transformer may cause serious problems like insulation failure. It is obvious that transformer needs a cooling system.

- Transformers can be divided in two types as (i) dry type transformers and (ii) oil immersed transformers.
- The method of cooling depends upon the (i) medium of cooling (ii) Type of circulation employed. They are abbreviated with standard notation such as
  Air – A, Gas – G, Synthetic oil – L, Mineral oil – O, Solid insulation – S, Water – W, Natural – N, Forced – F
- Coolant circulating inside the transformer comes in contact with winding and core, depending upon the method coolant transfers heat partially or fully to the tank walls from where it is dissipated to the surrounding medium.

Different Cooling Methods of Transformers are -

- For dry type transformers
  - ➢ Air Natural (AN)
  - ➢ Air Blast
- For oil immersed transformers
  - ➢ Oil Natural Air Natural (ONAN)
  - ➢ Oil Natural Air Forced (ONAF)
  - ➢ Oil Forced Air Forced (OFAF)
  - ➢ Oil Forced Water Forced (OFWF)

5.24.1 Cooling Methods For Dry Type Transformers

1. Air Natural (AN)

- This method of transformer cooling is generally used in small transformer upto 3 MVA). In this method the transformer is allowed to cool by natural air flow surrounding it.

2. Air Blast (AB)

- For transformers rated more than 3 MVA, cooling by natural air method is inadequate. In this method, air is forced on the core and windings with the help of fans or blowers. The air supply must be filtered to prevent the accumulation of dust particles in ventilation ducts. This method can be used for transformers upto 15 MVA.

5.24.2 Cooling Methods For Oil Immersed Transformers

1. Oil Natural Air Natural (ONAN)

- Air cooling is not sufficient and effective for medium and large size of transformer. Oil has main advantage of high heat conductivity and high co-efficient of volume expansion with temperature.
- Hence all transformers are immersed in oil and heat generated in windings and core are dissipated to the oil by conduction. In oil heat is transferred by convection. During convection heated oil transfers heat to the tank walls from where heat is taken away to the ambient air.

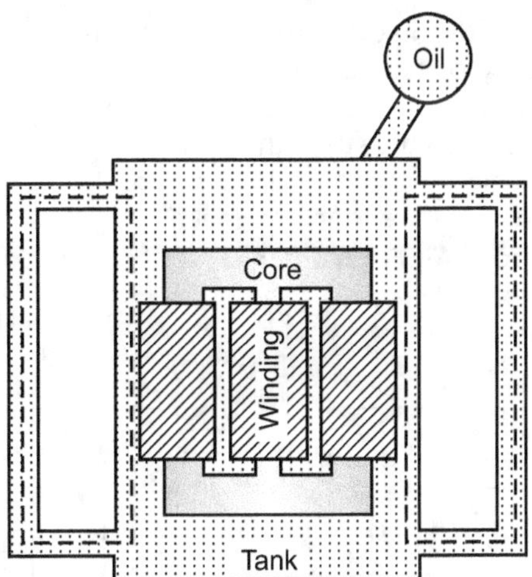

Fig. 5.24 : Oil natural air natural (ONAN) cooling

- During this process heated oil gets cooled and falls back to the bottom. Therefore natural thermal head is created which further transfer heat from heated part to the tank wall.
- Upto 30 kVA rating plain tank wall is sufficient to dissipate heat. Ratings higher than 30 kVA tank wall is increased by providing corrugations, fins, tubes and radiators.

2. Oil Natural Air Forced (ONAF)

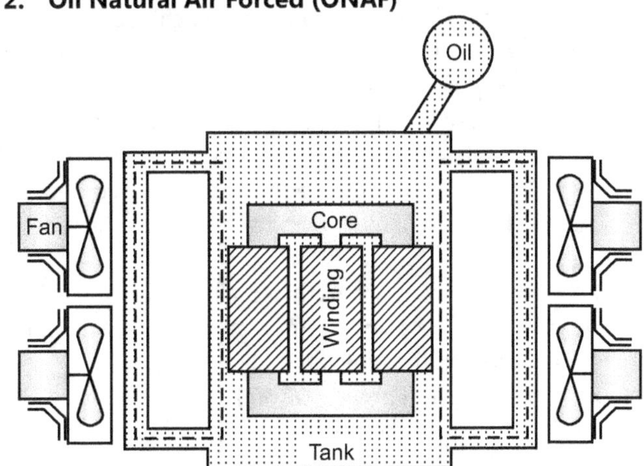

Fig. 5.25 : Oil natural air forced (ONAF) Cooling

- Transformers are immersed in oil where oil circulation under natural head transfers heat towards the tank wall.
- In this type of cooling techniques, tank is made hollow and air is blown through it to cool transformer parts. Heat removal from inner tank wall is increased by 5-6 times the conventional means.
- However normal way by air blast is to use radiator banks of corrugated tubes separated from tank and cooled by air blast produced by fan.

3. Oil Natural Water Forced (ONWF)

- In this type of cooling techniques, copper cooling coils are mounted above the transformer core but below the oil surface.

- Main disadvantage of this method is it employs water inside the oil tank and water is at higher head than oil. In case of water leakage, it gets mixed with oil and reduces dielectric strength of oil.
- Water inlet and outlet pipes are insulated in order to prevent moisture from ambient air.

4. Oil Forced Air Forced (OFAF)

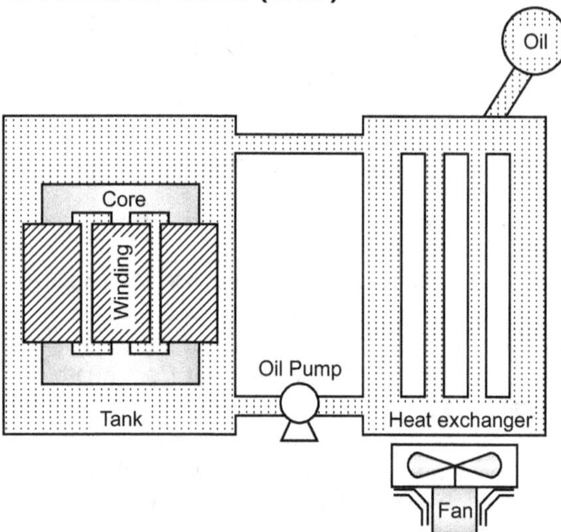

Fig. 5.26 : Oil forced air forced (OFAF) cooling

- In large transformer natural oil circulation is not enough to cool transformer, hence forced circulation through oil pump is carried out.
- A motor driven oil pump circulates oil from transform tank to external heat exchanger to cool the oil.
- In oil forced air forced arrangement oil is cooled by external heat exchanger using air blast produced by fan.
- Oil pumps and fans are not used all the time but, they are switched ON by temperature sensors when temperature exceeds the limit.

5. Oil Forced Water Forced (OFWF)

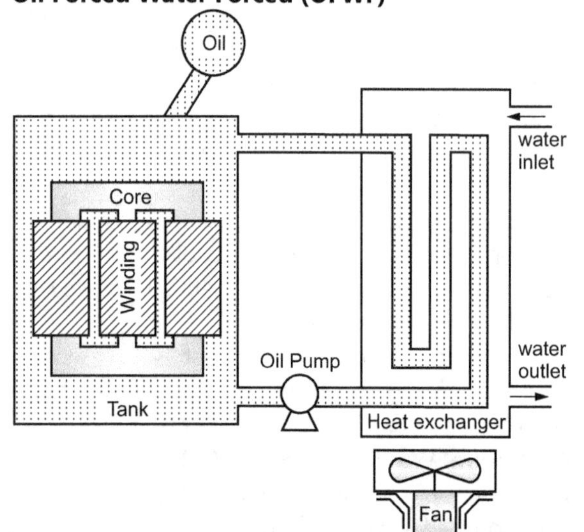

Fig. 5.27 : Oil forced water forced (OFWF) cooling

- In this oil forced water forced method oil is cooled in water heat exchanger. This method is best suited where cooling water has large head.
- The pressure of oil is kept higher than that of water so if any leakage that occurs will be from oil to water only.
- Hydroelectric generating stations has large water availability, hence transformer with OFWF cooling method is used there.

EXERCISE

1. What are the most important functions performed by transformers?

2. Compare core & shell type transformers.

3. What are distribution and power transformers?

4. List the value of K for different types of transformers.

5. What are square cores?

6. What are multi-stepped cores?

7. Mention the different paths of heat flow in transformer?

8. What are the various methods of cooling transformers?

9. Derive the output equation of single phase transformer

10. Derive the output equation of three phase transformer

11. Derive the equation for emf per turn of a transformer

12. Determine the core and yoke dimensions for a 250 kVA, 50Hz, single phase, core type transformer. Emf per turn = 1V, the window space factor = 0.33, current density 3A/mm and Bma = 1 .1 T. The distance between the centres of the square section core is twice the width of the core.

13. Obtain the expression for ratio of gross core area to circumscribing circle's area for (a) square core and (b) stepped core.

14. Derive the expression for magnetizing current of 1ϕ transformer.

15. Derive the expression for magnetizing volt ampere and magnetizing current of 3ϕ transformer.

16. Explain the various methods of cooling of transformer.

17. Obtain the expression for design of cooling tank with tubes and give the various clearance values.

COMPUTER AIDED DESIGN OF ELECTRICAL MACHINE

6.1 INTRODUCTION

Design of electrical machines mainly consists of, obtaining the dimensions of the various parts of the machine to suit given specifications, using available material economically and then to furnish these data to the manufacturer of the machine.

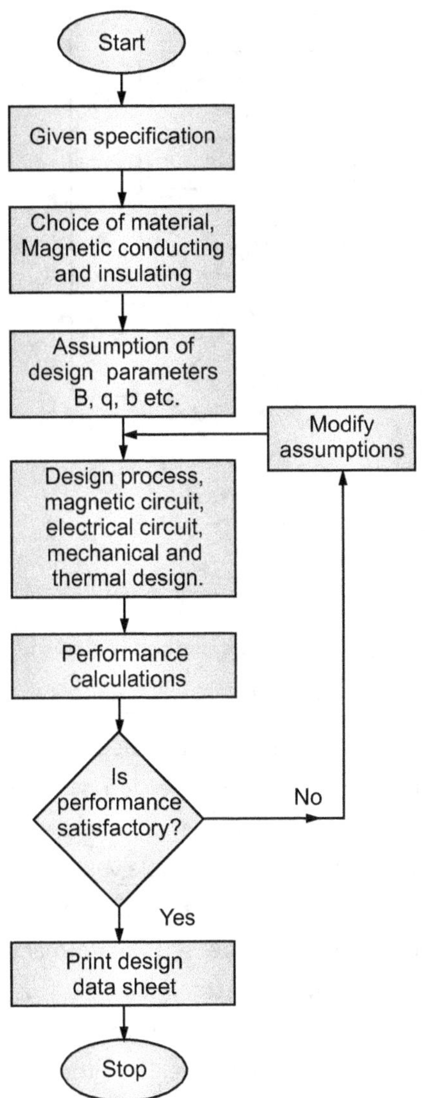

Fig. 6.1 : Design flow chart

The aim of the designer is to obtain.

(i) Lower cost

(ii) Smaller size

(iii) Wider temperature limit

(iv) Lower weight

(v) Better performance under no load and loaded conditions

(vi) Minimum losses

- The design procedure starts with the assumption of certain basic design parameters such as flux density, ampere conductors per metre or current density etc. From the given specifications of the machine and the above values of the design parameters, the dimensions of various parts of the machine are calculated. Based on these performance and temperature rise are worked out, which are then compared with the given specifications. If no satisfactory result is obtained, the basic assumed quanitites are suitably modified, till the result is upto the satisfaction.

- The design process is thus iterative as shown in design flow chart of Fig. 6.1 which is self explanatory. Normally, it takes many iterations to arrive at an optimal solution. The iterations require changes in values of variables, till both the cost and performance constraints are satisfied. This manual procedure of design (also known as hit and trial method) is tedious and time consuming. However, it can be done very quickly by use of digital computer. Hence, digital computer becomes a most valuable powerful tool in designing the electrical machines to satisfy the required performance of the customer.

- In the conventional design, we calculate the approximate temperature rise which has sufficient tolerance. For examples, in the case of power transformers, the approximate temperature rise of the windings above ambient may come out to be 50°C, but the actual temperature rise will be about 40°C. Thus, in the conventional design, the designer is unnecessarily forced to choose an insulating material having large temperature rise and thus making the design costlier.

- However, the use of computers in the design enables sophisticated calculations for the temperature rise, in the various parts of the machine. The exact hot spot temperature and its location can be determined with the use of computers. So, now for the same rating of the machine, an insulating material having lower temperature limit can be used and hence the initial cost of the machine can be reduced drastically.

6.2 ADVANTAGES OF COMPUTER AIDED DESIGN

The use of digital computers has completely revolutionized the field of design of electrical machines. Computer aided design eliminated the tedious and time consuming hand calculations, to achieve the desired result. The main advantages of computers in design of electrical machines are as follows :

- It has the capability to store large amount of design data.

- A large number of loops can be incorporated in the design programme.

- Arithmetic operations during design can be performed at approximately high speeds, which results in reducing the time.

- Highly accurate and reliable designs are obtained.

- Logical decisions can be taken, thereby saving the man hour of the design engineers.

- It is possible to optimize the design, with a reduction in overall cost and improvement in performance.

6.3 VARIOUS APPROACHES IN COMPUTER AIDED DESIGN

The two basic approaches commonly used for computer aided design of electrical machines are :

1. Analysis method

2. Synthesis method

6.3.1 Analysis Method

- In this method, the use of computer is made only for the purpose of analysis, leaving all the decision making to the designer. Fig. 6.2 gives the flow chart of the analysis method.

- In analysis method, the basic design parameters are chosen by the designer. These are fed to the computer as input data. The design details and performance is calculated by the computer. The performance is then examined by the designer and the he makes another choice of the input datas, if necessary, and the performance is recalculated. This process of design is repeated till the requirement are fully met with.

- Analysis method is the most simple method of computer aided design. As there are no logical decisions involved, hence the programming is also easier and time saving.

- However, optimization cannot be achieved with this approach.

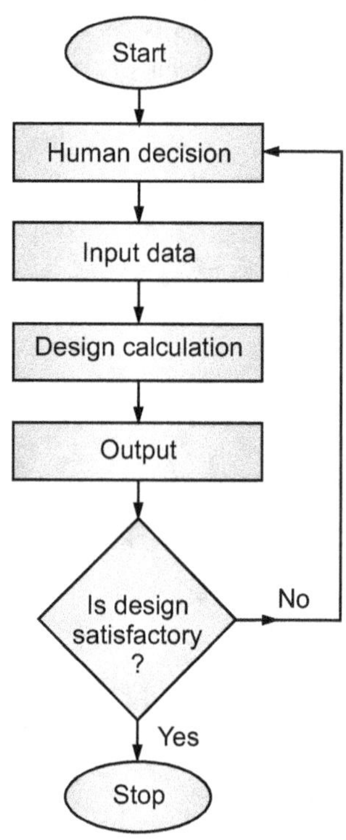

Fig. 6.2 : Analysis method of design

6.3.2 Synthesis Method

- In synthesis method, the logical decisions are taken by the computers instead of the designer. The required performance of the machine is fed as the input to the computer. It assumes the suitable values for the design parameters, calculates the performance and then compares with the desired performance. If the performance is unsatisfactory, than the computer itself adjusts the values of the design parameters and recalculates the performance. The process is repeated till a satisfactory performance is obtained.

- Flow chart illustrating this method has been given in Fig. 6.3. The main advantage of the synthesis method is that the optimization is possible. Most accurate and reliable designs can be obtained. However the programming in complex and time consuming.

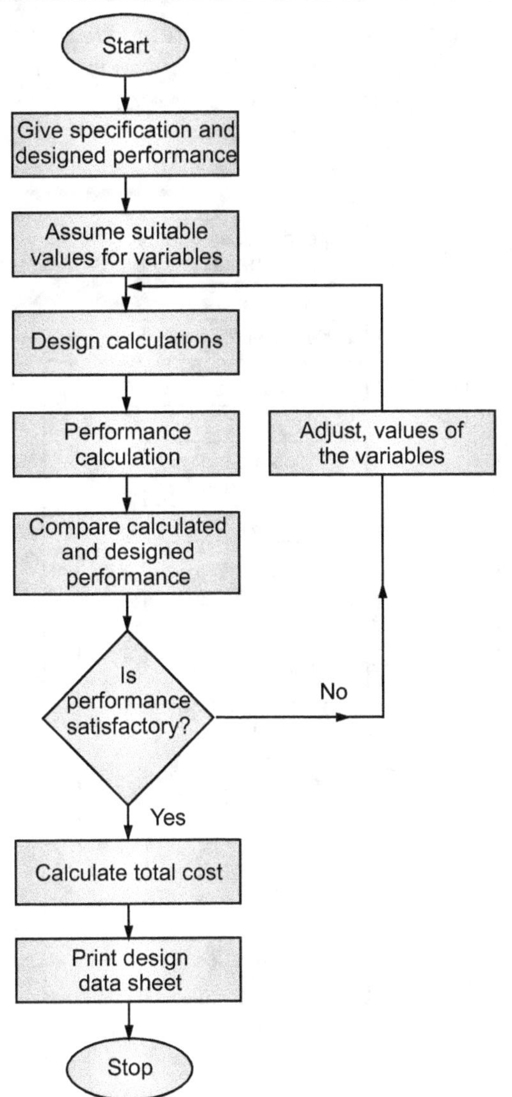

Fig. 6.3 : Synthesis method of design

6.4 COMPUTER AIDED DESIGN OF TRANSFORMER

- The philosophy followed in the computer aided design of transformers is the same as that adopted by an experienced designer. Similar to normal design, input data, such as KVA rating, voltage ratio of LV to HV winding, frequency etc. and the performance limitations like percentage regulation, percentage load losses temperature rise etc. are fed to the machine as input data. The computer selects suitable values to the design parameters, such as maximum flux density, current density, window space factor etc. based upon the rating of the machine to be designed. Using various design equations, computer calculates the complete design data and the performance characteristics. The performance is then compared by the computer with the specified limits imposed by the customer. If the performance is not up to the desired limitation, the initial assumptions of the design parameters are revised by the computer and the whole process in repeated, till a optimum design is obtained.

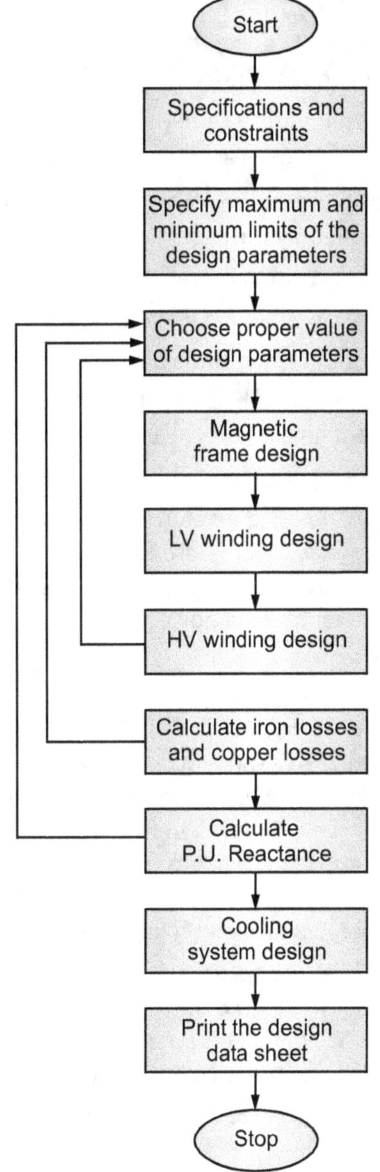

Fig. 6.4 : Flow chart for CAD of Transformer

Flow chart for the computer aided design of transformer has been shown in Fig. 6.4, which is more or less self explanatory. First of all input data including specifications and constraints, generally specified by the customer, are fed to the computer, such as –

(i) Continuous maximum rating of the transformer, KVA or MVA.

(ii) LV and HV line voltages, KV

(iii) Frequency, HZ

(iv) Number of phases

(v) LV and HV winding connection

(vi) Type of transformer

(vii) Percentage tapping

(viii) Percentage regulation

(ix) Percentage load losses

(x) Maximum temperature rise

(xi) Noise level, specially for large power transformer

Next the maximum or minimum limits of various design parameters are fed by the designer, for example

(i) Flux density in the core

(ii) Current densities in LV and HV winding

(iii) Maximum height of the transformer

(iv) Radial distance between the HV windings on two adjacent limbs.

(v) Window space factor.

- Now based upon the rated power, line voltage and winding connection, the proper values of these design parameters are chosen. Using the design parameters and the design equations, the magnetic frame in designed.

- Choosing proper values of the clearances between the windings and the thickness of the insulating cylinders, the LV and HV windings are designed using the conventional approach. If the HV winding on the two adjacent limbs overlap, then the design parameters are changed and the winding is redesigned is such a manner that now the HV windings do not overlap. Full load copper losses and iron losses are then calculated.

- The variation in copper losses is affected by increasing or decreasing the cross-sectional area of conductor of LV and HV winding. Iron losses are varied by changing the flux density in the core. Computer continuous the calculation with these changes till the full load copper losses and iron losses satisfy the specified limits. Percentage reactance, cooling system, the weight of core and copper are then computed and finally the detailed design data sheet is printed.

6.5 COMPUTER AIDED DESIGN OF ROTATING MACHINE

The computer aided design of rotating electrical machines is quite complicated, because of the following reasons.

1. There are a large number of independent variables. (ranging from 10 to 15 variables)

2. Some of the variables are discrete variables i.e. they can be varied in steps only. For example, number of stator slots, number of rotor slots etc.

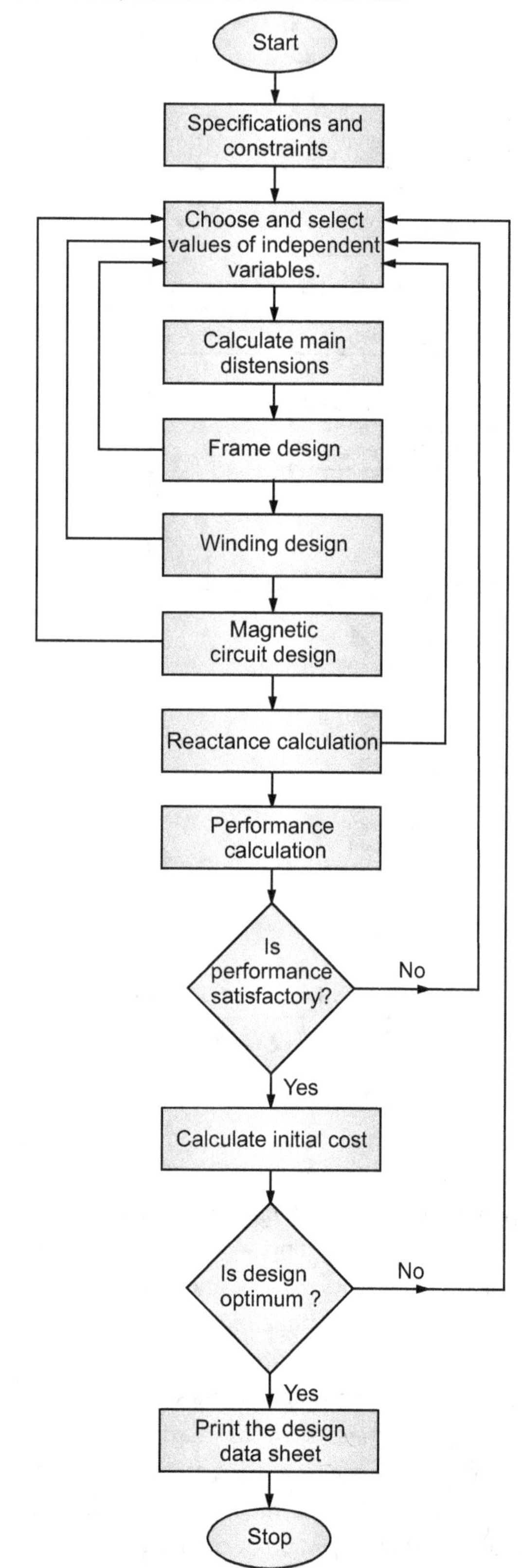

Fig. 6.5 : Flow chart of CAD of rotating electrical machines

The general flow chart for the computer aided design of rotating electrical machine is shown in Fig. 6.5. The given specifications and the independent variables must be chosen with almost care. There are a large number of variables in the actual design, of these, only a few have a significant effect on the performance. So, such variables are chosen as the independent variables, while the others are made constant for a particular design. In order to optimize the design, a large number of loops have been incorporated in the flow chart. If the performance is unsatisfactory, the independent variables are varied and the performance is recalculated till the satisfactory results are obtained. The design is then optimized for the minimum cost.

6.6 OPTIMIZATION OF DESIGN

A feasible design is one which satisfies all the given specifications and the required performance. But, a feasible design need not be an optimal one as regards the cost of active material, weight of active material, cost of energy wasted as losses and such other considerations. The additional optimization requirement can be achieved by formulating the design problem as a programming problem, in which the cost of active material or any such other criteria forms the objective function and the performance of the system form the constraints for the problem.

6.6.1 Design Problem Formulation

The most general design problem can be stated as :

Minimize (or maximize) the objective function $F(x_1, x_2, \ldots x_n)$ such that the variables $x_1, x_2, \ldots x_n$ satisfies the constraints $g_i(x_1, x_2, \ldots x_n)$ o, i = 1, 2, ... m.

The above statement can be easily explained by applying it for power transformers. For power transformers the independent variables $x_1, x_2, \ldots x_n$ may be chosen as follows :

1. Maximum magnetic flux density x_1 (wb/m^2)

2. Current density in HV winding x_2 (A/mm^2)

3. Current density in LV winding x_3 (A/mm^2)

4. Height of the winding x_4 (m)

5. Voltage per turn x_5 (V)

The objective function may be choosen from

1. Cost of the active material i.e. cost of copper winding and the cost of stampings.

2. Cost of energy wasted as losses.

3. Cost of the active material and cost of energy wasted as losses.

The constraint function $g(x_1, x_2, \ldots x_n)$ consists of the following :

1. Temperature rise of windings above ambient.

2. Temperature rise of top oil above ambient.

3. Percentage short circuit impedance.

4. Magnetic flux density in core.

5. Current density in winding.

6. Percentage efficiency at full load.

7. Percentage no load current.

6.6.2 Optimization Technique

The optimization technique can be broadly classified as follows :

1. Unconstrained optimization

2. Constrained optimization

- In unconstrained optimization, there is no upper and lower limits of the independent variables. The aim is to minimize (or maximize) the objective function only. For an unconstrained optimization problem, both gradient and non-gradient methods are available.

- In constrained optimization, the constraint are put on the independent variables. The aim is now to minimize (or maximize) the objective function such that the variables are in limits to the or constraint function are also satisfied. The constrained optimization problem can be solved using in interior or exterior penalty function technique or transformation of variables.

EXERCISE

1. Explain Various Approaches in Computer Aided Design

2. Explain the design procedure with flow chart in Computer Aided Design

3. Explain the analysis and synthesis method for CAD

4. Explain Computer Aided Design of Transformers

5. Computer Aided Design of Rotating Electrical Machines

6. Discuss the optimization of design with Problem Formulation & Optimization Technique

7. Discuss the advantages of Computer Aided Design

8. Explain the different optimization Technique of Computer Aided Design

MODEL QUESTION PAPERS FOR
End-Semester Examination

Paper- I

Time : 3 Hrs.   **Marks : 60**

Instructions to the candidates :

1.   Each Question carries 12 Marks.

2.   Attempt any five questions from the following.

3.   Illustrate your answers with neat sketches, diagram etc., wherever necessary.

4.   If some part or parameter is noticed to be missing, you may appropriately assume it and should mention it clearly.

1. **(a)** Explain limitations in design of electrical machines.   **[6]**

   (b) What is the significance of "Specifications" and "Standardization" in electrical machine Design?   **[6]**

2. **(a)** How would you calculate the size of wire of field rheostats for generators and motors.   **[6]**

   (b) A 1 kW, 250 V single element electric furnace is to employ a nicrome resistance wire operating at 1000°C. Estimate a suitable diameter and length of wire. Assume radiating efficiency = 1, emmisivity = 0.9 and and resistivity of wire 0.424 ohm- mt at 1000°C,The ambient temperature is 20°C.   **[6]**

3. **(a)** Develop a single layer lap winding diagram for a DC machine having 32 armature conductors and 4 poles. Mark the poles, draw the sequence diagram, indicate the position of the brushes and show the direction of induced emf.   **[6]**

   (b) Differentiate between front pitch and back pitch.   **[6]**

4. **(a)** What are the various modes of heat transfer from the inner most part of a machine to the surroundings.   **[6]**

   (b) Discuss the advantages of hydrogen as cooling medium as compared to air. What special precaution should be taken for hydrogen cooled alternator.   **[6]**

5. **(a)** Derive the output equation of three phase transformer.   **[6]**

   (b) Find the core and window areas required for a 1000 kVA, 6600/400 V, 50 Hz, single phase core type transformer. Assume a maximum flux density of 1.25 Wb/m^2 and a current density of 2.5 A/mm^2. Voltage per turn = 30 V. Window space factor = 0.32.   **[6]**

6. **(a)** Explain the analysis and synthesis method for CAD.   **[6]**

   (b) Computer Aided Design of Rotating Electrical Machines.   **[6]**

Paper- II

Time : 3 Hrs.   **Marks : 60**

Instructions to the candidates :

1. Each Question carries 12 Marks.

2. Attempt any five questions from the following.

3. Illustrate your answers with neat sketches, diagram etc., wherever necessary.

4. If some part or parameter is noticed to be missing, you may appropriately assume it and should mention it clearly.

1. **(a)** Briefly explain Biot-Savart's law as applicable in electric machine design.   **[6]**

    (b) Write short note on magnetic material.   **[6]**

2. **(a)** Derive the calculation of resistence steps of motor starters.   **[6]**

    (b) Describe a design procedure for series field winding.   **[6]**

3. Develop a double layer winding for a DC machine having 16 slots and 4 poles. Draw the sequence diagram and indicate the position of the brushes. Show the direction of induced emf and give equaliser connection.   **[12]**

4. **(a)** Derive an expression for the temperature rise curve for electrical machine   **[6]**

    (b) Write short note on hydrogen cooling.   **[6]**

5. **(a)** The ratio of flux to full load mmf in a 400 kVA, 50 Hz, single phase core type power transformer is 2.4×10^{-6}. Calculate the net iron area and the window area of the transformer. Maximum flux density in the core is 1.3 Wb/m^2, current density 2.7 A/mm^2 and window space factor 0.26. Also calculate the full load mmf.   **[6]**

    (b) Obtain the expression for design of cooling tank with tubes and give the various clearance values.   **[6]**

6. **(a)** Explain the design procedure with flow chart in Computer Aided Design.   **[6]**

    (b) Explain the different optimization Technique of Computer Aided Design.   **[6]**

✠ ✠ ✠

www.ingramcontent.com/pod-product-compliance
Lightning Source LLC
Chambersburg PA
CBHW060118120726
48003CB00009B/2691